graphic products

revised edition

Tristram Shepard ■ Andrew Loft

First published 1996 by
Stanley Thornes (Publishers) Ltd

This edition published in 2001 by:
Nelson Thornes Ltd
Delta Place
27 Bath Road
CHELTENHAM
GL53 7TH
United Kingdom

A catalogue record for this book is available from the British Library.

ISBN 0 7487 6081 4

02 03 04 05 / 10 9 8 7 6 5 4 3

Designed and typeset by Carla Turchini
Artwork by Andrew Loft, Hardlines, Mark Dunn
Picture research by John Bailey, axonimages.com
Printed and bound in Italy by Canale

The authors would particularly like to thank Amanda White of Film Education for her suggestions for the Movie Marketing project, and Ralph Ardill, Sandy Grice and Sarah Macdonald of Imagination for the Product Launch Case Study material and Anthony Wheaton of Guernsey Press for technical information and details of paper and printing processes.

Thanks are also due to Russ Jones and Jet Mayor for their contributions to the revised edition.

The authors would like to acknowledge the influence of the following books, and particularly recommend them as sources of further information for teachers and pupils:

The Colour Eye, Robert Cumming and Tom Porter, BBC Books, 1990
About Printing, Ivor Powell and Alan Jamieson, Hobsons Publishing, 1989
About Packaging, Hobsons Publishing 1993
The Creation of a Carton, BPCC Taylowe Ltd, 1993
Printing Effects, Wyne Robinson, Macdonald, 1991
Presentation Techniques, Dick Powell, Macdonald, 1990
Green Design, Dorothy Mackenzie, Laurence King, 1991
How to Design Trademarks and Logos, John Murphy and Michael Rowe, Phaidon, 1988
How to Design Grids, Alan Swann, Phaidon, 1989
Colour Proof Correction, David Bann and John Gargan, Phaidon, 1990
Mastering Manufacturing, Gordon Mair, Macmillan, 1993

Sources of Further Information

The British Printing Industries Federation, 11 Bedford Row, London WC1R 4DX
Film Education, 27-31 Charing Cross Road, London WC2H 0AU, www.filmeducation.org
Film Education may be able to supply a video on the marketing of the film Judge Dredd which would provide useful further stimulus material for 'Movie Marketing'.

The publishers are grateful to the following for permission to reproduce photographs or other illustrative material:

ACCO-UK: pp.112–13 (far left)
Adobe: front cover and title page (lower right), pp. 40–1, 42–3 (left, upper right), 118–19 (left)
AKA: pp. 118–19 (top, right)
Bentley: pp. 106–7, 108–9, 110–11, 128–9
Berol Karisma: pp. 64–5 (lower right)
BHS: pp. 46–7 (left)
Body Shop International: pp. 68–9 (top right)
Bridgeman Art Library: pp. 6 (bottom right – The Procession of Pope Gregory I (c. 540–604) through Rome in 590, from the litanies of a Book of Hours, French, early 16th century (vellum) by French School (16th century) courtesy of the Trustees of Sir John Soane's Museum, London/Bridgeman Art Library), 7 (bottom left – Interior of printing works, from 'Royal Verses on the Conception', 1537 (vellum) by French School (16th century) Bibliotheque Nationale, Paris, France/Index/Bridgeman Art Library), 34 (top left and right – The Harbour and the Quays at Port-en-Bessin, 1888 by George Pierre Seurat (1859–91) Minneapolis Institute of Arts, MN, USA/Lauros-Giraudon/Bridgeman Art Library)
British Gas: pp. 46–7 (top right)
British Heart Foundation (registered charity number 22591): p.54
Countryside Commission: pp. 46–7 (left)
Dairy Crest: p. 8 (top)
Denford Computerised Machines: p.13 (top left)
Department of Transport: pp. 46–7 (left)
Disabled Living Foundation: p.54
Ford: pp. 100–101
Getty Telegraph Colour Library: p.9 (bottom – Gone Loco)
Habitat UK: pp. 42–3
Help the Aged: p. 54
Hemera Technologies: pp. 6 (top), 18, 20, 21, 22, 23, 24, 25, 26, 54 (top), 144 (top)
Hulton Getty: p.7 (bottom right – A Hudson)
Imagination Limited: pp. 36–7, 138–9, 140–1, 142–3
JVC: pp. 13 (bottom left), 144 (bottom)
Kodak: pp. 114–15 (upper and lower far left)
Le Shuttle: pp. 46–7 (top right)
Martyn Chillmaid: cover and title page (upper left with thanks to Arjo Wiggins, bottom centre), pp. 13 (top right), 152 (with thanks to Arjo Wiggins), 34–5 (lower), 46–7 (lower right with thanks to Marcello's Ristorante & Pizzeria, Q Multimedia Ltd., Ariss Design, Collie Books, Wells Fargo Bank, Imagination Ltd.), 58–9, 66–7 (top right and centre right with thanks to Coca Cola), 68–9 (bottom right), 70–1, 72–3, 74–5, 76–7 (with thanks to Elida Gibbs, BrightReason Restaurant Group and Tetley), 78–9, 84–5 (upper far left, right), 104–5 (upper, lower left)
MCA: pp. 28–9 (lower right), 30–1 (lower right) – © 1996 U.C.S. and Amblin. TM Harvey. All rights reserved.
Nichol Beauty Products: pp. 74–5 (lower far left)
Oxfam: pp. 46–7 (top right), p. 54
Panasonic UK: p.13 (bottom right)
Popperfotos: p. 7 (top left)
The Post Office: pp. 28–9 (lower left – © is vested in the Post Office. This material is reproduced by permission of the Post Office.)
PrimaryUK: p.145
Robert Opie: pp. 98–9
Rolls-Royce PLC: pp. 82–3 (lower), 120–21 (lower left)
The Ronald Grant Archive: pp. 28–9 (upper right), 30–1 (bottom left, centre top, top right)
Science Pictures Limited: pp. 72–3 (with thanks to Nichol Beauty Products), 94–5, 92–3 (centre bottom)
Simon Phillips: front cover and title page (lower left), pp.36–7 (Bostik is a registered trademark of Bostik Ltd and is used with permission), 48–9, 90–1, 94–5, 112–13 (top right, upper and lower right), 114–15 (centre), 116–17, 122–3 (bottom right), 142–3 (right)
Sky TV: pp. 46–7 (left)
Sony: pp. 6, 17, 96–7 (lower left), 124–5
Toni The Maldive Lady, Maldive Travel, London: pp. 44–5
Walls Birdseye: pp. 84–5 (left)
Wicksteed Leisure Limited: pp. 150–1
ZEFA: pp. 66–7 (bottom right), 68–9 (left, upper right, lower right), 82–3 (upper), 120–1

Contents

Introduction

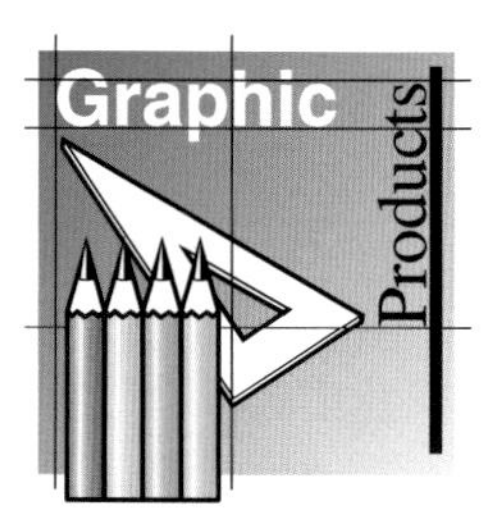

Welcome to* Design & Make It: Graphic Products. *This book has been written to support you as you work through your GCSE course in Design and Technology. It will help guide you through the important stages of your coursework, and assist your preparation for the final examination paper.

How to Use this Book

There are a number of different ways in which you might use this book. For example:

- ▷ using the Project Guide at the start of the book to check that you are covering and documenting your coursework in the way the examiners will be looking for
- ▷ following one or more of the four main design and make projects, undertaking a selection of the focused activities as you go
- ▷ referring to the specific pages which cover the areas of knowledge and understanding (e.g. Talking Typography, Into Print, Drawing in Three Dimensions) while doing any design and make project
- ▷ attempting the sample examination paper questions, basing your answers on the material covered in the preceding pages
- ▷ considering one of the three Project Suggestions at the end of the book as possible final projects.

The Set Projects

If you follow the projects (Movie Marketing, Mix and Match, Keeping In Touch and Product Launch) as presented in the book you will be covering key areas of your examination syllabus as you go. At the same time as developing your coursework skills you will prepare yourself for the final examination paper.

In each of the projects the development of one possible solution has been used as an example. You can base your work on this solution, but do try to come up with ideas of your own. This is particularly important if you want to achieve higher marks and a better final grade.

Long or short?
If you are following a short course, check with your teacher which sections of the book you do not need to cover.

Making it
Whatever your project, remember that the final realisation is particularly important. It is not enough to just hand in your development folio. You must have separate pieces of final artwork (preferably mounted on card), and/or presentation models or working products. The quality of this final realisation is particularly important, and counts for a high proportion of the marks.

During your course you will need to develop your technical skills in using graphic materials and equipment. This is something you can't do just by reading a book! The best way is to watch carefully as different techniques and procedures are demonstrated to you, and practise them as often as possible.

■ ACTIVITY

Make sure that as part of your project folio you include evidence of having completed a number of short-term focused practical tasks, as suggested in the Activity sections

IN YOUR PROJECT

Use the In Your Project paragraphs to help you think about how you could apply the content of the page to your current work.

KEY POINTS

Use the Key Points paragraphs to revise from when preparing for the final examination paper.

Beyond GCSE
Printing and publishing is the UK's eighth largest industry. Electronic publishing in particular is an exciting growth area which will be needed to serve our rapidly developing global information and communication technologies. There is a wide range of further courses at various levels which you might like to find out more about.

Design Matters

What is Design and Technology, how has it changed, and why is it important?

As you develop your ideas for graphic products you will often need to make important decisions about the social, moral, cultural and environmental impact of your product.

How does Design and Technology Affect Our Lives?

Technology helps extend our natural capabilities. For example, it enables us to:

- travel further and faster
- send and receive messages across the world in an instant
- keep us warm in winter and cool in summer.

Designers help make new and existing technologies easy and more pleasant for people to use - they make them look and feel fun and fashionable, logical and safe to use. They also work out how to make them easy to produce in quantity, and cheap to manufacture and sell.

So Design and Technology is about improving people's lives by designing and making the things they need and want. But different people have different needs: what is beneficial to one person can cause a problem for someone else, or create undesirable damage to the environment.

A new design might enable someone to do something quicker, easier and cheaper, but might cause widespread unemployment or urban decay. It could also have a harmful impact on the delicate balance of nature.

As you develop your design ideas you will often need to make important decisions about the social, moral, cultural and environmental impact of your product.

A brief history of graphic products

The first graphic products were probably cave-paintings. In Greek and Roman times, quills, sticks and brushes were used to mark onto clay tablets, animal skins, papyrus or paper.

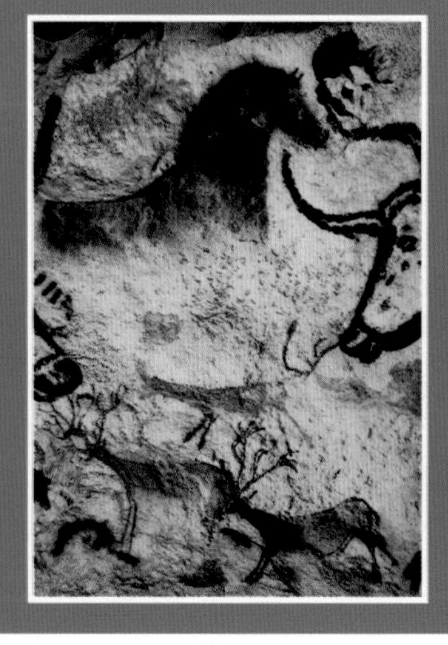

Eighth–Twelfth Century
The Chinese printed text and pictures using hand-carved wooden blocks. In Europe monks produced illuminated books. These were hand-drawn and took a long time to do. Only a small number were made, and they could only be afforded by the very rich.

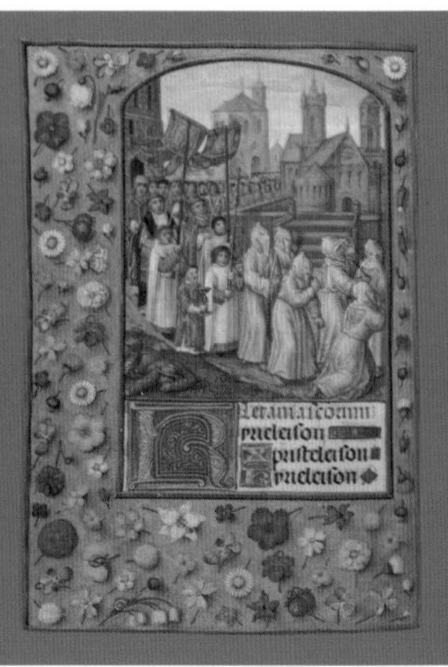

Environmental Issues

We all need to be aware of the amount and use of energy and resources, as on our earth these are finite. The more we use up natural resources without replacing them, the fewer there are for future generations. Graphic products use paper and petro-chemicals that use non-replaceable energy and contributes to global warming. Is all that packaging really necessary?

Social Needs

Good design can help bring people together. Designers need to be careful about creating products that might have the effect of isolating someone, or making them more vulnerable to crime in some way.

Cultural Awareness

People from different cultures think and behave in different ways. What is acceptable to one culture may be confusing or insulting to another. For example, in Britain most families use kettles to boil drinking water. This is not the case in France where many use a pan. Colours and certain shapes can have very different meanings across the world.

Moral Issues

Sometimes designers are asked to develop products that can cause harm to people or animals. Would you be willing to create a graphic product that aimed to sell cosmetics that you knew were derived from animals that had been slaughtered for this purpose?

Fifteenth Century
The movable type printing press was invented by Johanns Gutenburg in Germany in 1455. The type was still hand-made but could be re-arranged. This made the process of printing faster and cheaper, so many copies could be printed.

Nineteenth Century
Newspapers started to be produced using cylindrical presses printing onto long rolls of paper. These were very cheap per copy.

Industrial Matters

Good design involves creating something that works well and is satisfying to use. But to be successful a product also needs to be commercially viable.

Professional Designers

Different designers specialise in creating different sorts of products. Professional designers usually specialise in different areas. For example:

- ▷ 3D Design
- ▷ 2D Design
- ▷ Fashion and textiles
- ▷ Architecture.

This book is particularly concerned with Graphic Products. These can include all the above areas.

What is a Graphic Product?

Many graphic products are printed or reproduced in some way onto a flat surface, for example, advertisements, CD inlays, road signs, stamps, etc. They are sometimes applied to a three-dimensional object. Three-dimensional models that are created to communicate ideas about a proposed design, e.g. a model of a theatre set or a new household product, are also known as graphic products.

A graphic product outcome (i.e. what you end up making) is likely to be a prototype or appearance model of an artefact that could be manufactured or reproduced in quantity. Some outcomes may be complete products, e.g. musical greetings cards, children's activity books. Others may not be fully 'working', but their finish, presentation and detail should realistically represent a 'real' artefact.

It should be remembered that the purpose of the outcome is to communicate a design concept to a potential client, manufacturer or purchaser.

Twentieth Century
High-quality reproduction of images and colourwork developed. During the last quarter of the century digital computer technology revolutionises the production process. There is much greater flexibility in design and layout, greatly increased speed and the ability to send text and images across the world in seconds.

New developments in material technology enable text and Images to be reproduced on a much wider range of surfaces.

When this piece of black card is placed on a warm heat source the image of the car appears.

Design for Profit

Products are designed and made to make life easier and more enjoyable, or to make a task or activity more efficient. However, along the way the people who create these products need to make a profit. The designer needs to do more than satisfy the needs of the market, and to consider the sorts of issues described on page 7. They must also take into account the needs of the clients, manufacturers and retailers. The aim is to design and make products that are successful from everyone's point of view.

Designers:
- agree a brief with a client
- keep a notebook or log of all work done with dates so that time spent can be justified at the end of the project
- check that an identified need is real by examining the market for the product
- keep users' needs in mind at all times
- check existing ideas. Many designers re-style existing products to meet new markets because of changes in fashion, age, environment, materials, new technologies, etc
- consider social, environmental and moral implications
- consider legal requirements
- set limits to the project to guide its development (design specification)
- produce workable ideas based on a thorough understanding of the brief
- design safe solutions
- suggest materials and production techniques after considering how many products are to be made
- produce working drawings for manufacturers to follow.

Manufacturers need to:
- make a profit on the products produced
- agree and set making limits for the product (manufacturing specification)
- develop marketing strategies
- understand and use appropriate production systems
- reduce parts and assembly time
- reduce labour and material costs
- apply safe working procedures to make safe products
- test products against specifications before distribution
- produce consistent results (quality assurance) by using quality control procedures
- understand and use product distribution systems
- be aware of legislation and consumer rights
- assume legal responsibility for product problems or failures.

Retailers:
- need to make a profit on the products sold
- consider the market for the product
- give consumers what they want, when they want it, at an acceptable price
- take account of consumers' legal rights
- take consumer complaints seriously
- continually review new products
- put in place a system to review and replace stock levels.

Consumers/users expect the product to:
- do the job it was designed for
- give pleasure in use
- have aesthetic appeal
- be safe for its purpose
- be of acceptable quality
- last for a reasonable lifetime
- offer value for money.

Clients:
- identify a need or opportunity and tell a designer what they want a product to do and who it's for (the brief)
- consider the possible market for the idea
- organise people, time and resources and raise finance for the project.

Twenty first Century

During the first decade of the current century developments in electronic communication technologies will continue to develop rapidly in new and exciting ways.

The computer and television screen will become more integrated. Everything will become much more closely linked together. You will start to have much more choice and control about what you want to see and hear.

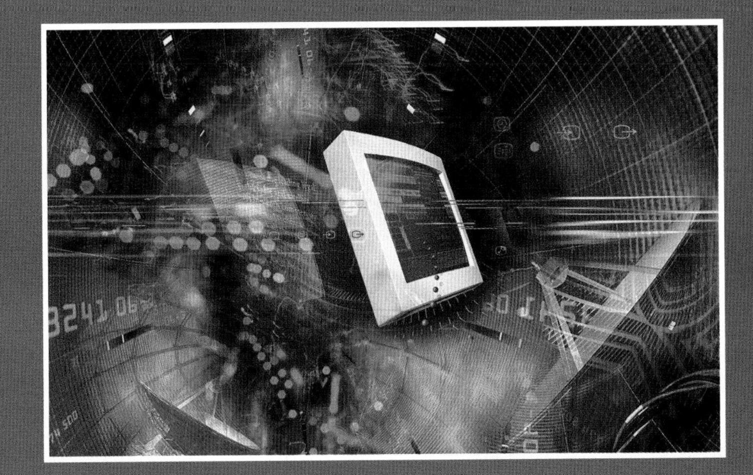

Using ICT in Graphic Products (1)

ICT (Information and Communication technology) is widely used in the design and production of graphic products, as you will discover. You can considerably enhance your GCSE coursework with the effective use of ICT.

Using ICT in your Work

To gain credit for using ICT you need to know when it is best to use a computer to help with your work. Sometimes it is easier to use ICT to help with parts of your coursework than to do it another way. On other occasions it can be far easier to write some notes on a piece of paper than use a computer – this saves you time and helps you to do the job more effectively.

Following are some ideas showing you how using ICT could enhance your coursework. Some can be used at more than one stage. You do not have to use all them!

Identifying the Problem

The internet could be used to search manufacturers' and retailers' web-sites for new products, indicating new product trends.

Project Planning

A time chart can be produced showing the duration of the project and what you hope to achieve at each stage using a word processor or DTP program. Some programs allow you to produce a Gantt chart (see page 83).

Investigation

- ▷ A questionnaire can be produced using a word processor or DTP. Results from a survey can be presented using a spreadsheet as a variety of graphs and charts.
- ▷ Use a digital camera to record visits and existing products
- ▷ The internet can be used to perform literature searches and communicate with other people around the world via e-mail.

Search engines

To help you find the information you need on the internet you can use a search engine. A search engine is a web-site that allows you to type in keywords for a specific subject. It then scans the internet for web sites that match what you are looking for. Here are the addresses of some popular search engines:

www.excite.co.uk
www.yahoo.co.uk
www.netscape.com
www.hotbot.co.uk
www.msn.co.uk
www.searchtheweb.com

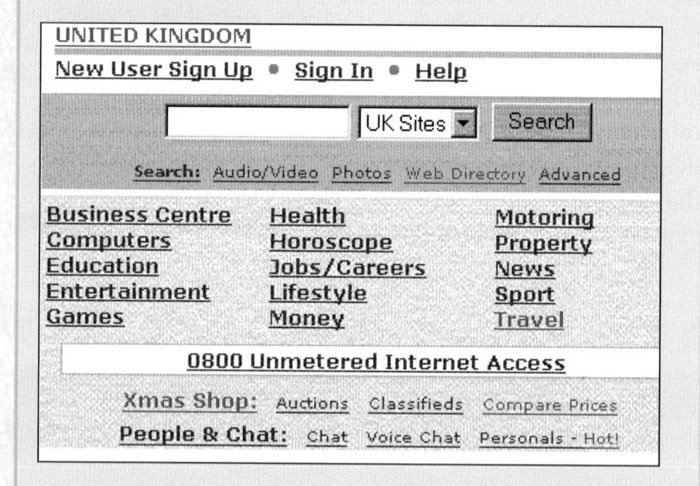

E-mail

E-mail is a fast method of communicating with other people around the world. Text, photographs and computer files can be attached and shared. Some web-sites have e-mail addresses - you could try to contact experts to see if they could help with your coursework. It is important to be as specific as you possibly can, as these experts may be very busy people.

Specification

A design or product specification can be written in a word processor. Visual images of the product, diagrams and other illustrations could also be added, and easily modified at a later date.

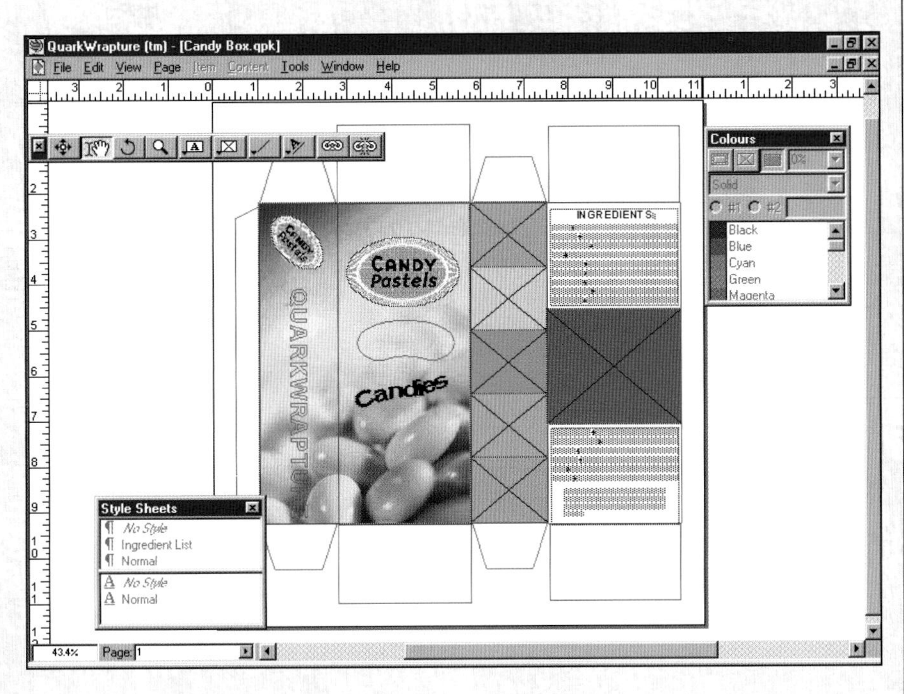

Developing Ideas

- ▷ Ideas for your product could be produced using a graphics program, DTP or 3D design package. Colour variations can be applied to product drawings to test a design on its intended market before production.
- ▷ Likely costs of new products can be modelled using a spreadsheet. Different component costs can be modelled quickly and easily allowing you to see the consequences of your design ideas.

The cups shown above were designed and shaded (rendered) using a CAD program.

Final Ideas and Production

- ▷ A document showing the specification, images, production method and components can be word-processed.
- ▷ Parts lists and the costs of materials can be calculated and displayed using spreadsheets.
- ▷ A detailed flow production diagram could be produced using a DTP program. Images could also be added to show important stages.
- ▷ Digital images can be used in the production plan as a guide to show how the product should be assembled or to indicate its colour.
- ▷ Pre-programmed CAM equipment could be used to replicate manufacture, (see next page).
- ▷ Packaging nets can be produces using graphics package or by using templates, e.g. **www.dtonline.org.** Scanned or digital photographs can then be placed within the templates to produce instant packaging designs.

Project Presentation

- ▷ Use graphics packages to prepare text and visual material for presentation panels. Charts showing numerical data can be quickly produced using a spreadsheet
- ▷ Use a presentation package, such as *PowerPoint* to communicate the main features of your design.

Using ICT in Graphic Products (2)

The use of Computer-aided Design and Computer-aided Manufacture (CAD-CAM) are particularly appropriate in Graphic Products and will help you enhance your coursework.

CAD/CAM

Computer Aided Design (CAD) and Computer Aided Manufacture (CAM) are terms used for a range of different ICT applications that are used to help in the process of designing and making products.

Computer systems contain three main elements – a series of **inputs** that are **transformed** into **outputs**.

CAD is a computer aided system for creating, modifying and communicating ideas for a product or components of a product.

CAM is a broad term used when several manufacturing processes are carried out at one time aided by a computer. These may include process control, planning, monitoring and controlling production.

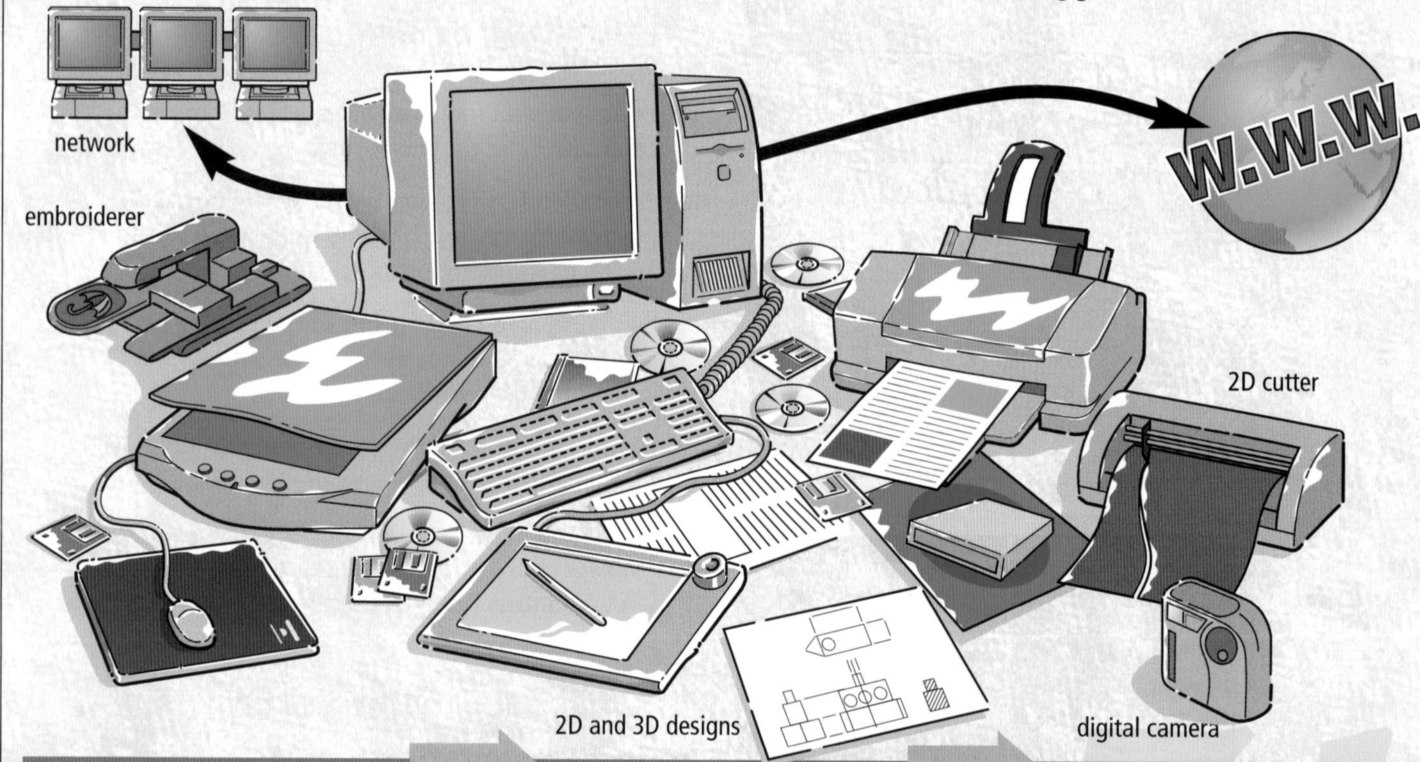

INPUTS	TRANSFORMATION	OUTPUT
Text, drawings and photographs are put into the system using devices such as keyboards, drawing tablets and scanners.	The computer's processor and program change the information as required, providing feedback to the designer via the monitor, and responding to on-screen tools and control devices.	The final design is turned into an output through a colour or black and white printer, or as instructions to a computer numerically controlled (CNC) machine to create a 3D object.

From Design to Making: a Compact Disc Player

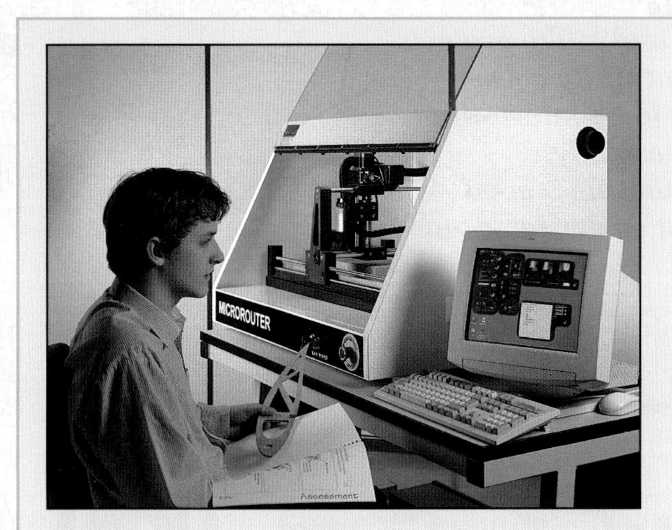

The designer develops the shape and form of the CD player casing on the computer screen, using appropriate software. It can be created as a:

- dimensioned technical drawing
- pictorial image that can be rendered and rotated (a 'virtual' prototype)

The final design is converted into numerical data that can be read by a CNC machine. This data can either be stored on disc or sent directly to a CNC router or milling machine.
The CNC router or miller, reads the data and cuts the CD player from a block of modelling material. This solid model is sealed and painted to represent the real player, and decorated with self-adhesive vinyl 'stickers' to indicate control buttons, product name, etc.

Any necessary changes to the final design are made on screen, and sent to another CNC machine that will produce either a vacuum former mould or an injection moulding tool. CAD data can be communicated to a manufacturing site well away from where the designing was done. The machine can be anywhere – in a different room, a different building, or even in another country!

ICT is also used to create:

- a secure packaging for the CD player
- the surface decoration of the package
- sales promotional literature
- a point of sale display

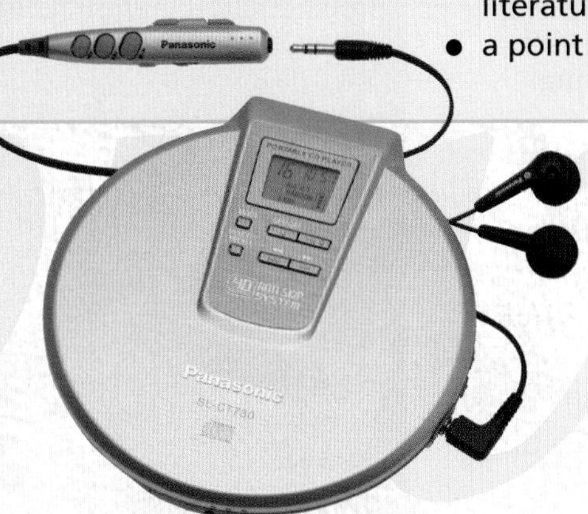

In industry, ICT and CAD-CAM are typically used to produce:

2D outputs

- working drawings – plans and maps
- designs for packages – nets
- labelling – packages, containers
- surface decoration on packaging
- books, posters and information sheets
- electronic circuits manufacture
- fabric & embroidery designs
- ceramic decorations\signs in public spaces
- advertising and promotional materials
- designs for passing on for 3D forming

3D outputs

- patterns & moulds for manufacturing processes
- quick prototyping
- point of sale displays
- engraving
- scanning 3D objects

Choosing and Starting Projects

Identifying suitable design and make projects for yourself is not easy. A carefully chosen project is much more likely to be interesting and easier to complete successfully. Investing time and effort choosing a good project makes progress a lot easier later on.

the home

energy

the natural environment

the high street

transport

communications

clothing

leisure

security

food

health

education

Project Feasibility Studies

Make a start by making a list of:

- ▷ potential local situations/environments you could visit where you could do some research into the sort of things people there might need (e.g. a local playgroup, a small business, a derelict piece of land, etc.)
- ▷ people you know outside school who might be able to help by providing information, access and / or advice.

The next stage is to get up and get going. Arrange to visit some of the situations you've listed. Choose the ones which you would be interested in finding out some more about. Make initial contact with the people you know, and get them interested in helping you. Tell them about your D&T course, and your project.

You should visit each possible situation to identify people's needs and spot opportunities to help solve their problems with new graphic products and systems.

With a bit of luck, after you've done the above you should have a number of ideas about possible projects.

Try to identify what the possible outcomes of your projects might be – not what the final design would be, but the form your final realisation might take, e.g. an artwork dummy, a scale model, a series of plans and elevations, etc. Think carefully about the following:

- ▷ Might it be expensive or difficult to reproduce or make?
- ▷ Do you have access to the tools and materials which would be required?
- ▷ Will you be able to find out how it could be printed or manufactured?
- ▷ Does the success of the project depend on important information you might not be able to get in the time available?
- ▷ Are there good opportunities for you to use ICT?

starting points

There are different ways in which you might be able to start a project. Your teacher may have:

- told you exactly what you are required to design
- given you a range of possible design tasks for you to choose from
- left it up to you to suggest a possible project.

If you have been given a specific task to complete you can go straight on to page 16.

If you are about to follow one of the main units in this book, you should go straight to the first page of the task.

If you have been given a number of possible tasks to choose from you should go straight to the section entitled 'Making your Choice'.

However, if you need to begin by making some decisions about which will be best task for you, then the first stage is to undertake some feasibility studies.

Making Your Choice

For each of your possible projects work through pages 16 and 17 (Project Investigation) and try planning out a programme of research.

Look back over your starting questions and sources of information:

- ▷ Could you only think up one or two areas for further research?
- ▷ Did you find it difficult identifying a range of sources of information?

If this has been the case, then maybe it is not going to prove to be a very worthwhile project.

Ideally, what you're looking for is a project which:

- ▷ is for a nearby situation you can easily use for research and testing
- ▷ you can get some good expert advice about
- ▷ provides good opportunities for you to do a range of research activities and an expected outcome which will make it possible for you to make and test a prototype
- ▷ will not be too difficult to finally realise
- ▷ shows a good use of ICT
- ▷ is suitable for reproduction or manufacture.

Finally, one of the most important things is that you feel interested and enthusiastic about the project!

don't forget...

in my design folder

- ✓ *My project is to design a...*
- ✓ *I am particularly interested in...*
- ✓ *I have made a very good contact with...*
- ✓ *My prototype can be tested by...*
- ✓ *My final outcome will include...*
- ✓ *I could use ICT to...*

Project Investigation

You will need to find out as much as you can about the people and the situation you are designing for. To do this you will need to identify a number of different sources of information to research into.

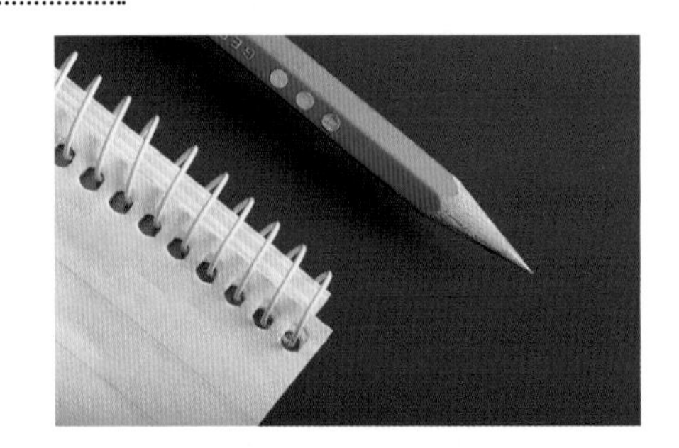

Starting Questions

Make a list of questions you will have to find answers to.

You should find the following prompts useful:

Why...? When...? Where...?

What...? How many...? How often...?

How much...?

Sources of Information

Next, carefully consider and write down the potential sources of information you might be able to use in order to discover the answers to your starting questions.

Work through the research methods on the next page. Be sure to give specific answers as far as possible (i.e. name names).

Across your research you will need to aim to obtain overall a mixture of:

- ▷ factual information: e.g. size, shape, weight, cost, speed, etc.
- ▷ information which will be a matter of opinion: i.e. what people think and feel about things, their likes and dislikes, what they find important, pleasing, frustrating, etc.

don't forget...

Write down what you need to find out more about, and how you could obtain the information.

You need to use a range of sources of information (e.g. user research, existing solutions, expert opinion, information search). The more methods you use, the more marks you will get.

When you undertake the research, remember to record what you discover.

Make sure your research work is clearly and attractively presented.

in my design folder

- ✓ The key things I need to find out about are...
- ✓ The research methods I am going to use are...
- ✓ I will be talking to the following people about my project...
- ✓ I will need to have it all completed by...
- ✓ I will use ICT to...

Research Methods

User Research

Which people could you observe and consult who are directly involved in the situation? To what extent do you consider that you will be able to find out about:

- the things they do
- the way in which they do them.

As well as asking individuals, you could also undertake a small survey or questionnaire.

User Trips

How can you record your own impressions of the situation? Are there any relevant activities you could try out for yourself to gain first-hand experience? Do you have any recollections of any previous similar experiences you have had?

Site Study

In what ways could you document the environment in which the situation is? Which of the following will be relevant?

- Historical and geographical factors.
- Sociological, economic, political information.
- Location.
- Layout, facilities.
- Sizes and spaces.
- Atmosphere – light, colour, texture.
- The surrounding environment.

Similar Situations

Do you know of any other comparative circumstances in which people are in similar situations, and which might help provide insight and ideas?

Expert Opinion

Are there any people you know of who could give you expert professional advice on any aspects of the situation? If you don't know immediately of anyone, how might you set about finding somebody?

Information Search

Has any information about the situation, or a similar situation, been documented already in books, magazines, TV programmes, the internet, or CD-ROM, etc.? If you don't already know that such information exists, where could you go to look for it? Don't forget to consider the possibility of using information stored on a computer database.

In Conclusion

When most of your investigation work has been completed you will need to draw a series of conclusions from what you have discovered. What have you learnt about the following things:

- ▷ What sort of people are likely to be using the product?
- ▷ Where and when will they be using it?
- ▷ What particular features will it need to have?
- ▷ How many should be printed or made?

in my design folder

- ✓ From my research I found out...
- ✓ I have discovered that...
- ✓ My conclusions are that...
- ✓ I have kept my research relevant by...
- ✓ I found ICT helpful when...

From Design Specification to Product Specification

A design specification is a series of statements that describe the possibilities and restrictions of the product. A product specification includes details about the features and appearance of the final design, together with its materials, components and manufacturing processes.

Writing a Design Specification

The **design specification** is a very important document. It describes the things about the design which are fixed and also defines the things which you are free to change.

The conclusions from your research should form the starting point for you specification. For example, if in your conclusions you wrote:

'From the measurements I made I discovered that the best length for an arm-rest would be between 250 and 400 mm.'

In the specification you would simply write:

'The arm-rest should be between 250 and 400 mm long.'

The contents of the specification will vary according to the particular product you are designing, but on the opposite page is a checklist of aspects to consider. Don't be surprised if the specification is quite lengthy. It could easily contain 20 or more statements.

Fixing it

Some statements in the specification will be very specific, e.g.: *'The toy must be red.'*

Other statements may be very open ended, e.g.: *'The toy can be any shape or size.'*

Most will come somewhere in between, e.g.: *'The toy should be based on a vehicle of some sort and be mechanically or electronically powered.'*

In this way the statements make it clear what is already fixed (e.g. the colour), and what development is required through experimentation, testing, and modification (e.g. shape, size, vehicle-type and method of propulsion).

Writing a Product Specification

After you have fully developed your product you will need to write a final more detailed **product specification**. This time the precise statements about the materials, components and manufacturing processes will help ensure that the manufacturer is able to make a repeatable, consistent product.

Your final product will need to be evaluated against your design specification to see how closely you have been able to meet its requirements, and against your product specification to see if you have made it correctly.

don't forget...

You might find it helpful to start to rough out the design specification first, and then tackle the conclusions to your research. Working backwards, a sentence in your conclusion might need to read:

'From my survey, I discovered that young children are particularly attracted by bright primary colours.'

It's a good idea to use a word processor to write the specification. After you've written the design specification new information may come to light. If it will improve the final product, you can always change any of the statements.

Make sure you include as much numerical data as possible in your design specification. Try to provide data for anything which can be measured, such as size, weight, quantity, time and temperature.

Specification Checklist

The following checklist is for general guidance. Not all topics will apply to your project. You may need to explore some of these topics further during your product development.

Use and performance

Write down the main purpose of the product – what it is intended to do. Also define any other particular requirements, such as speed of operation, special features, accessories, etc. Ergonomic information is important here.

Size and weight

The minimum and maximum size and weight will be influenced by things such as the components needed and the situation the product will be used and kept in.

Generally the smaller and lighter something is the less material it will use, reducing the production costs. Smaller items can be more difficult to make, however, increasing the production costs.

Appearance

What shapes, colours and textures will be most suitable for the type of person who is likely to use the product? Remember that different people like different things.

These decisions will have an important influence on the materials and manufacturing processes, and are also crucial to ensure final sales.

Safety

A product needs to conform to all the relevant safety standards.

- Which of them will apply to your design?
- How might the product be mis-used in a potentially dangerous way?
- What warning instructions and labels need to be provided?

Conforming to the regulations can increase production costs significantly, but is an area that cannot be compromised.

Manufacturing cost

This is concerned with establishing the maximum total manufacturing cost which will allow the product to be sold at a price the consumer or client is likely to pay.

The specification needs to include details of:

- the total number of units likely to be made
- the rate of production and, if appropriate
- the size of batches.

Maintenance

Products which are virtually maintenance free are more expensive to produce.

- How frequently will different parts of the product need to be maintained?
- How easy does this need to be?

Life expectancy

The durability of the product has a great influence on the quantity of materials and components and manufacturing process which will need to be used.

How long should the product remain in working order, providing it is used with reasonable care?

Environmental requirements

In your specification you will need to take into account how your product can be made in the most environmentally friendly way. You might decide to:

- specify maximum amounts of some materials
- avoid a particular material because it can't be easily recycled
- state the use of a specific manufacturing process because it consumes less energy.

Other areas

Other statements you might need to make might cover special requirements such as transportation and packaging.

in my design folder

✓ *My design will need to...*
✓ *The requirements of the people who will use it are...*
✓ *It will also need to do the following...*
✓ *It will be no larger than...*
✓ *It will be no smaller than...*
✓ *Its maximum weight can be...*
✓ *It should not be lighter than...*
✓ *The shapes, colours and textures should...*
✓ *The design will need to conform to the following safety requirements...*
✓ *The number to be printed or made is...*
✓ *The following parts of the product should be easily replaceable...*
✓ *To reduce wastage and pollution it will be necessary to ensure that...*

Generating and Developing Ideas

When you start designing you need lots of ideas – as many as possible, however crazy they might seem. Then you need to start to narrow things down a bit by working in more detail and evaluating what you are doing.

Work towards making at least one prototype to test some specific features of your design out. Record the results and continue to refine your ideas as much as you can. Sorting out the final details often requires lots of ideas too.

First Thoughts

Start by exploring possibilities at a very general level. Spend time doing some of the following:

▷ Brainstorming, using key words and phrases or questions which relate to the problem.
▷ Completing spider-diagrams which map out a series of ideas.
▷ Using random word or object-association to spark off new directions.
▷ Thinking up some good analogies to the situation (i.e. What is it like?).
▷ Work from an existing solution by changing some of the elements.
▷ Experimenting with some materials.

Continue doing this until you have at least two or three possible approaches to consider. Make sure they are all completely different, and not just a variation on one idea.

Go back to your design specification. Which of your approaches are closest to the statements you made? Make a decision about which idea to take further, and write down the reasons for your choice.

As you work through this section it is important to remember the following sequence when considering potential solutions:

- record a number of different possibilities
- consider and evaluate each idea
- select one approach as the best course of action, stating why.

There are lots of different drawing techniques which you can use to help you explore and develop your ideas, such as plans, elevations, sections, axonometrics, perspectives, etc.
Try to use as rich a mixture of them as possible. At this stage they should really be 'rough', rather than 'formal' (i.e. drawn with a ruler). Colour is most useful for highlighting interesting ideas.

don't forget...

As usual, it is essential to record all your ideas and thoughts.

Much of your work, particularly early on, will be in the form of notes. These need to be neat enough for the examiner to be able to read.

Drawings on their own do not reveal very well what you had in mind, or whether you thought it was a good idea or not. Words on their own suggest that you are not thinking visually enough. Aim to use both sketches and words.

As you develop your ideas, make sure you are considering the following:

- Design – aesthetics, ergonomics, marketing potential, etc.
- User requirements – functions and features.
- Technical viability – if it could be made.
- Manufacturing potential – how it could be made in quantity.
- Environmental concerns – if it can be reused, recycled, etc.?

Second Thoughts

Working on paper, begin to develop your ideas in more detail. Remember to use a range of drawing techniques, such as plans, elevations, 3D sketches, as well as words and numbers to help you model your ideas. Wherever possible, consider using a CAD program. Don't forget to print out the various stages you work through, or at least make sure you keep your own copy on disk of the screens you generate.

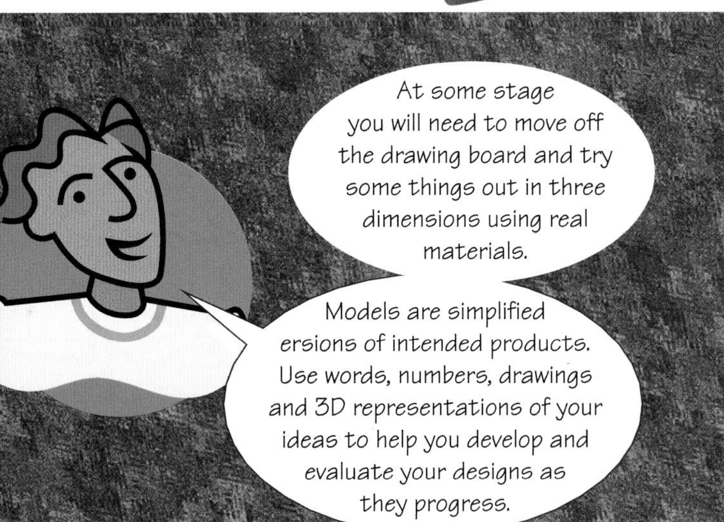

Planning and Making Prototypes

At some stage you will be in a position to bring your ideas into sharper focus by making some form of mock-up or prototype. Think carefully about exactly what aspect of your idea you want to test out and about the sort of model which will be most appropriate.

Whatever the form of your final outcome, the prototype might need to be:

- ▷ two-dimensional
- ▷ three-dimensional
- ▷ at a different scale
- ▷ made using different materials.

Try to devise some objective tests to carry out on your prototype involving measuring something. Don't rely just on people's opinions. Write up the circumstances in which the tests were undertaken, and record your results.

Write down some clear statements about:

- ▷ what you wanted the prototype to test
- ▷ the experimental conditions
- ▷ what you discovered
- ▷ what decisions you took about your design as a result.

Following your first prototype you may decide to modify it in some way and test it again, or maybe make a second, improved version from scratch. Make sure you keep all the prototypes you make, and ideally take photographs of them being tested perhaps using a digital camera.

Sometimes you will need to go back to review the decisions you made earlier, and on other occasions you may need to jump ahead for a while to explore new directions or to focus down on a particular detail. Make sure you have worked at both a general and a detailed level.

in my design folder

- ✓ *I chose this idea because...*
- ✓ *I developed this aspect of my design by considering...*
- ✓ *To evaluate my ideas I decided to make a prototype which...*
- ✓ *The way I tested my prototype was to...*
- ✓ *What I discovered was...*
- ✓ *As a result I decided to change...*
- ✓ *I used ICT to...*

Planning the Making and the Manufacturing

The final realisation is very important. It presents your proposed design solution rather than the process you used to develop it. Careful planning is essential. You will also need to be able to explain how your product could be reproduced or manufactured in quantity.

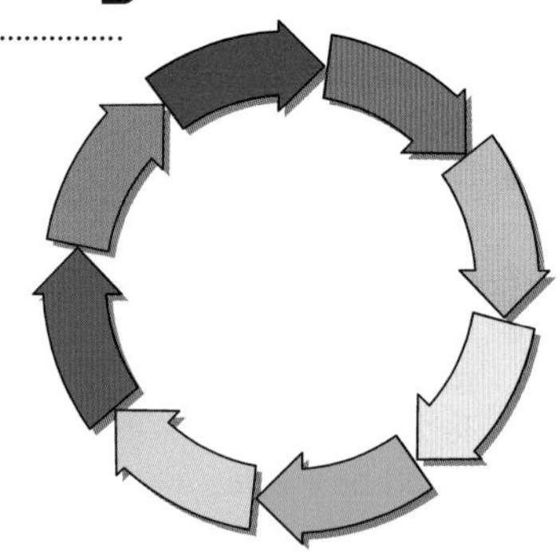

How many?
What you have designed should be suitable for manufacture or reproduction. You should discuss with your teacher how many items you should attempt to make. This is likely to depend on the complexity of your design and the materials and facilities available in your workshops. It may be that you only make one item, but also provide a clear account of how a quantity of them could be produced.

keeping a record

Write up a diary record of the progress you made while making. Try to include references to:

- things you did to ensure safety
- the appropriate use of materials
- minimising wastage
- choosing tools
- practising making first
- checking that what you are making is accurate enough to work
- asking experts (including teachers) for advice explaining why you had to change your original plan for making.

A Plan of Action

Before you start planning you will need to ensure that you have a final artwork rough of your design. If it is a 3D product you will need an orthographic drawing (see page 106): this will need to include all dimensions and details of the materials to be used. Ideally there should also be written and drawn instructions which would enable someone else to be able to make up the design from your plans.

Next work out a production flow chart as follows.

1. List the order in which you will prepare the final artwork or make the main parts of the product. Include as much detail as you can (see pages 103 to 115).
2. Divide the list up into a number of main stages, e.g. gathering materials and components, preparing (i.e. marking out, cutting), assembling, finishing.
3. Identify series of operations which might be done in parallel.
4. Indicate the time scale involved on an hourly, daily and weekly basis.

Think carefully about what reprographic tools are available in your school. Photography, photocopiers and computers could all save you a considerable amount of time.

If you are making 3D items you should consider the use of templates and jigs to speed things up.

Other possibilities include the use of moulds or setting up a simple CAM system to produce identical components (see page 62).

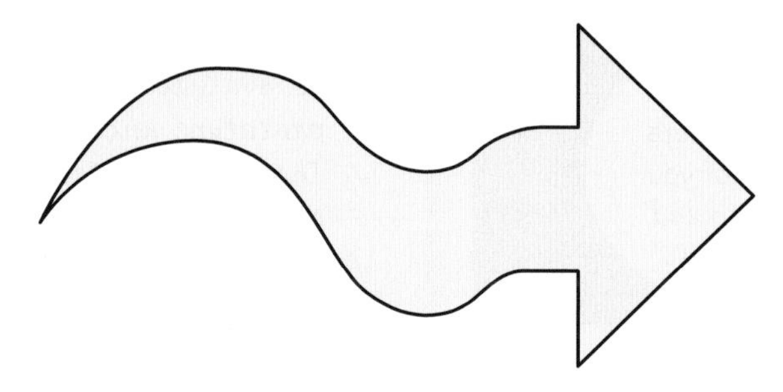

don't forget...

Quality Counts

As your making proceeds you will need to check frequently that your work is of acceptable quality. How accurately will you need to work? What tolerances will be acceptable? (See page 86.) How can you judge the quality of the finish? What might cause a product to be rejected?

If you are making a number of identical items you should try and work out ways of checking the quality through a sampling process (see page 87).

Manufacturing matters

Try asking the following questions about the way your design might be made in quantity:

- What work operation is being carried out, and why? What alternatives might there be?
- Where is the operation done, and why? Where else might it be carried out?
- When is it done, and why? When else might it be undertaken?
- Who carries it out, and why? Who else might do it?
- How is it undertaken, and why, How else might it be done?

Remember that manufacturing is not just about making things. It is also about making them better by making them:

- simpler
- quicker
- cheaper
- more efficient
- less damaging to the environment.

Making

While you are in the process of making you must ensure that the tools and materials you are using are the correct ones. Pay particular attention to safety instructions and guidelines.

Try to ensure that you have a finished item at the end, even if it involves simplifying what you do.

Aim to produce something which is made and finished as accurately as possible. If necessary you may need to develop and practise certain skills beforehand.

Planning for Manufacture

Try to explain how your product would be manufactured in quantity. Work through the following stages:

1. Determine which type of production will be most suitable, depending on the number to be printed or made.
2. Break up the production process into its major parts and identify the various sub-assemblies.
3. Consider where jigs, templates and moulding processes could be used. Where could 2D or 3D CAM be effectively used?
4. Make a list of the total number of components and volume of raw material needed for the production run.
5. Identify which parts will be made by the company and which will need to be bought in ready-made from outside suppliers.
6. Draw up a production schedule which details the manufacturing process to ensure that the materials and components will be available exactly where and when needed. How should the workforce and workspace be arranged?
7. Decide how quality control systems will be applied to produce the degree of accuracy required.
8. Determine health and safety issues and plan to minimise risks.
9. Calculate the manufacturing cost of the product.
10. Review the design of the product and manufacturing process to see if costs can be reduced

More information on all these topics can be found on pages 81 to 91.

in my design folder

- ✓ I planned the following sequence of making...
- ✓ I had to change my plan to account for...
- ✓ I used the following equipment and processes...
- ✓ I paid particular attention to safety by...
- ✓ I monitored the quality of my product by...

Testing and Evaluation

You will need to find out how successful your final design solution is. How well does it match the design specification? How well have you worked? What would you do differently if you had another chance?

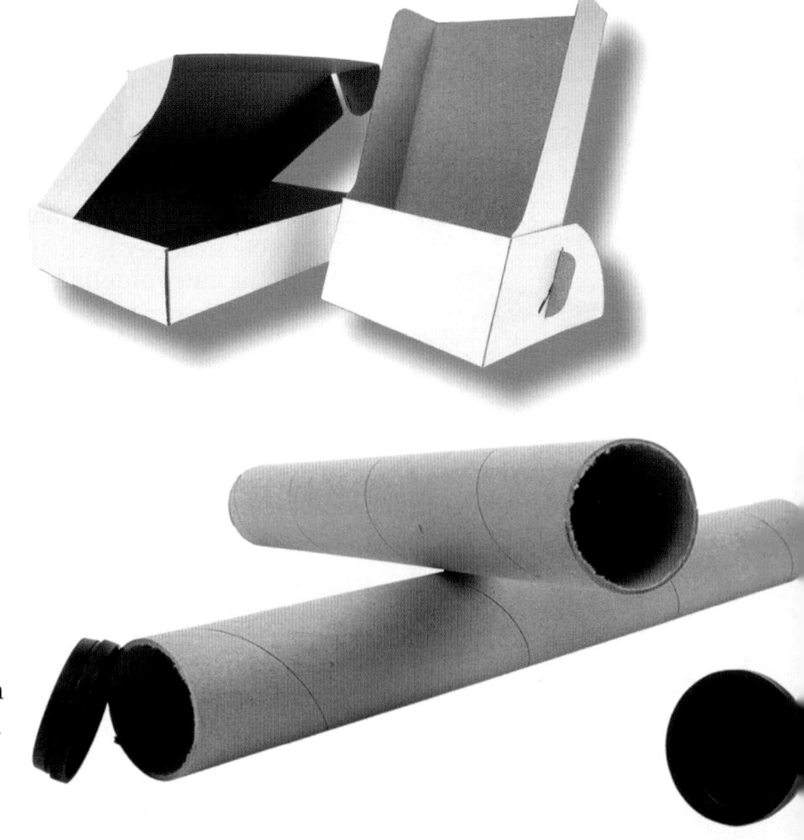

Testing the Final Solution

To find out how successful your design is you will need to test it out. Some of the ways in which you might do this are by:

- trying it out yourself
- asking other people to use it
- asking experts what they think about it.

As well as recording people's thoughts, observations and opinions, try to obtain some data: how many times it worked, over what periods of time, within what performance limits, etc?

To help you decide what to test, you should look back to the statements in your design specification, and focus on the most important ones. If for example the specification stated that a three-year-old child must be able to operate it, try and find out how many can. If it must be a colour which would appeal to young children, devise a way of finding out what age ranges it does appeal to.

You need to provide evidence to show that you have tested your final design out in some way. Try to ensure that your findings relate directly to the statements in your original specification. Include as much information and detail as you can.

don't forget...

Final Evaluation

There are two things you need to discuss in the final evaluation: the quality of the product you have designed, and the process you went through while designing it.

The product

How successful is your final design? Comment on:

- ▷ how it compares with your original intentions as stated in your design specification
- ▷ how well it solves the original problem
- ▷ the materials you used
- ▷ what it looks like
- ▷ how well it works
- ▷ what a potential user said
- ▷ what experts said
- ▷ whether it could be reproduced or manufactured cheaply enough in quantity to make a profit
- ▷ the effective use of ICT to assist reproduction or manufacture
- ▷ the extent to which it meets the requirements of the client, manufacturer and the retailer.
- ▷ the ways in which it could be improved.

Justify your evaluation by including references to what happened when you tested it.

The process

How well have you worked? Imagine you suddenly had more time, or were able to start again, and consider:

- ▷ Which aspects of your investigation, design development work and making would you try to improve, or approach in a different way?
- ▷ What did you leave to the last minute, or spend too much time on?
- ▷ Which parts are you most pleased with, and why?
- ▷ How well did you make the final realisation?
- ▷ How effective was your use of ICT? How did it enhance your work?

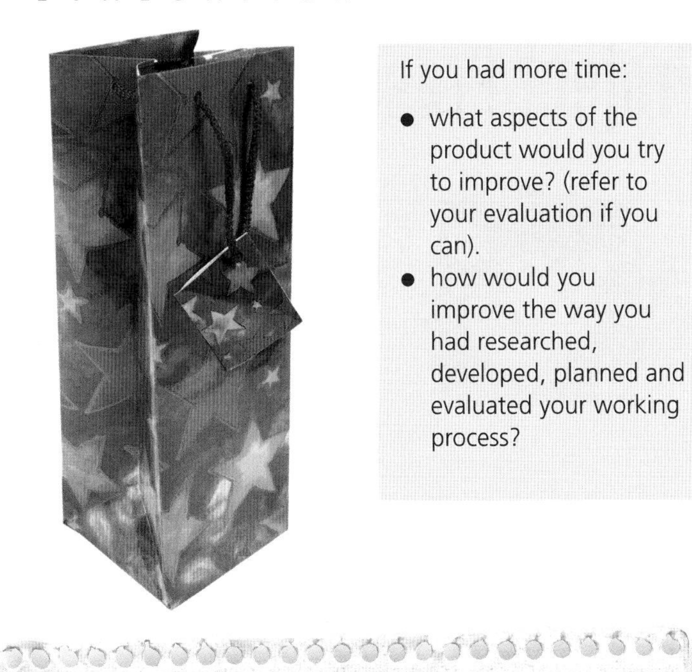

If you had more time:

- what aspects of the product would you try to improve? (refer to your evaluation if you can).
- how would you improve the way you had researched, developed, planned and evaluated your working process?

in my design folder

What do you think you have learnt through doing the project?

- ✓ *Comparison of my final product specification with my design specification showed that...*
- ✓ *The people I showed my ideas (drawings and final product) to said...*
- ✓ *I was able to try my design out by...*
- ✓ *I discovered that...*
- ✓ *I could improve it by...*
- ✓ *I didn't do enough research into...*
- ✓ *I spent too long on...*
- ✓ *I should have spent more time on...*
- ✓ *The best aspect is...*
- ✓ *I have learnt a lot about...*

Try to identify a mixture of good and bad points about your final proposal and method of working. You will gain credit for being able to demonstrate that you are aware of weaknesses in what you have designed and the way that you have designed it.

If people have been critical of aspects of your design, do you agree with them? Explain your response.

Project Presentation

The way you present your project work is extremely important. Remember you won't be there to explain it all when it's being assessed! You need to make it as easy as possible for an examiner to see and understand what you have done.

Telling the Story

All your investigation and development work needs to be handed in at the end, as well as what you have made. Your design folder needs to tell the story of the project. Each section should lead on from the next, and clearly show what happened next, and explain why. Section titles and individual page titles can help considerably. Try to ensure you have at least one A3 sheet which covers each of the headings shown opposite.

There is no single way in which you must present your work, but the following suggestions are all highly recommended:

- ▷ Securely bind all the pages together in some way. Use staples or treasury tags. There is no need to buy an expensive folder.
- ▷ Add a cover with a title and an appropriate illustration.
- ▷ Make it clear which the main sections are.
- ▷ Add in titles or running headings to each sheet to indicate what aspect of the design you were considering at that particular point in the project.

Remember to include evidence of ICT work and other Key Skills. Carefully check through your folder and correct any spelling and punctuation mistakes.

don't forget...

Presenting your Design Project Sheets

- ▷ Always work on standard-size paper, either A3 or A4.
- ▷ Aim to have a good mixture of text and visual images. These could be produced by hand, or on a computer.
- ▷ You might like to design a special border to use on each sheet.
- ▷ Include as many different types of illustration as possible.
- ▷ When using photographs, use a small amount of adhesive applied evenly all the way round the edge to secure them to your design folder sheet. A small amount of paper adhesive applied evenly all the way round the edge is the best way of fixing them on.
- ▷ Think carefully about the lettering for titles, and don't just put them anywhere and anyhow. Try to choose a height and width of lettering which will be well balanced on the whole page. If the title is too big or boldly coloured it may dominate the sheet. If it is thin or light it might not be noticed.

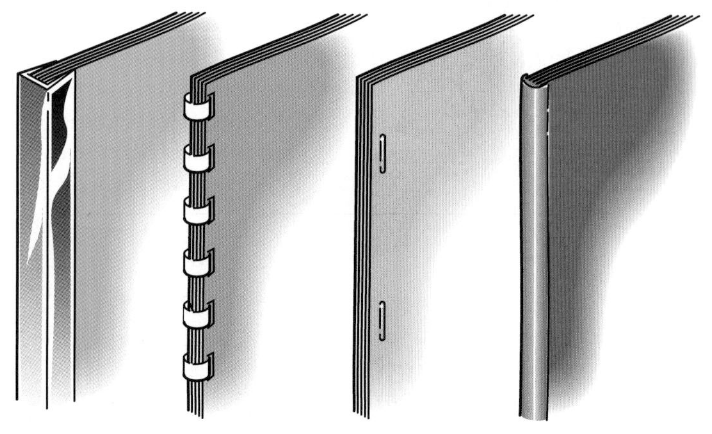

Binding methods

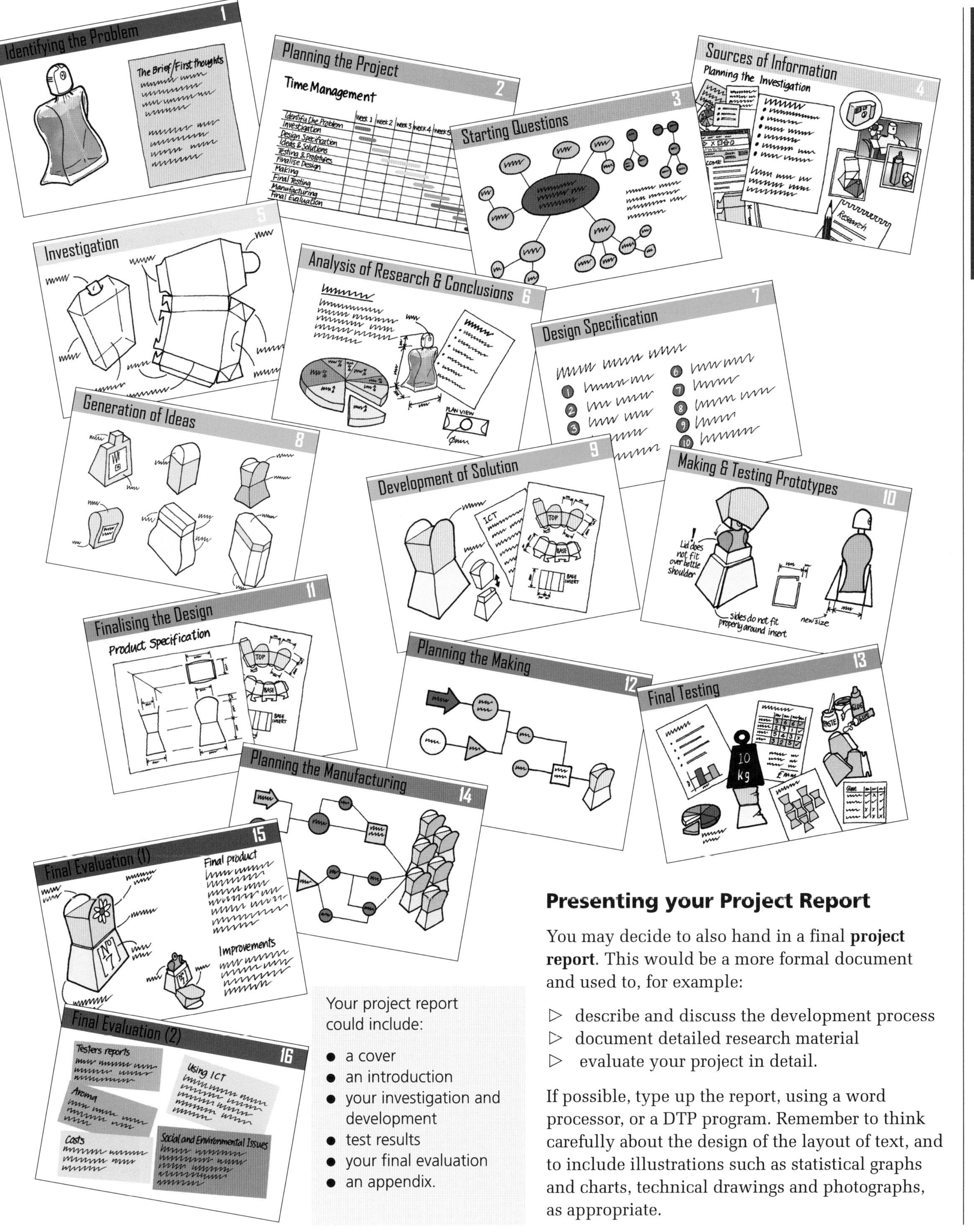

Your project report could include:

- a cover
- an introduction
- your investigation and development
- test results
- your final evaluation
- an appendix.

Presenting your Project Report

You may decide to also hand in a final **project report**. This would be a more formal document and used to, for example:

▷ describe and discuss the development process
▷ document detailed research material
▷ evaluate your project in detail.

If possible, type up the report, using a word processor, or a DTP program. Remember to think carefully about the design of the layout of text, and to include illustrations such as statistical graphs and charts, technical drawings and photographs, as appropriate.

Project One: Introduction

A major production company are about to launch a new movie in the UK. The main focus of the publicity will be a poster. This will be supported by a wide range of tie-ins, such as T-shirts, video packaging, magazine competitions, and so on.

You are asked to design:

- ▷ ***the main poster***
- ▷ ***a promotional point-of-sale display unit for the cinema foyer***
- ▷ ***a range of merchandising which uses the key image and motif.***

Designing Point-of-Sale Displays (page 50)

Graphic Identity (page 46)

WWW.

Visit **www.filmeducation.org** to find out more about films and how they are marketed.

The first silent movies were shown in 1895. It was not until 1927 that 'talkies' were invented and the 1940s when colour was used for major films.

Today movies are big business. Millions are spent making a film which will only be shown in cinemas for a few weeks after its launch.

Promoting a film through posters, promotional events and merchandising are essential to its commercial success.

If your school has a copy, now is a good time to watch the 'Marketing Judge Dredd' Case Study video.

The Task

The production company have provided you with the following description of the content of the film:

```
EMERALD BAY

At the end of the day, will green turn to
grey?

An abandoned boat on scenic Emerald Bay is
the start of this tense, explosive
adventure. Matthew Black and Magenta
Hirari find themselves underwater - well
over their heads and out of their depth.
Marooned on a small island, they must
somehow prevent this tranquil paradise
with its rare, endangered species becoming
a lifeless nuclear disaster area.

Can Matt and Magenta's computer wizardry
stop the mysterious Dr Dream setting off a
sequence of events which would upset the
balance of nature forever?
```

You will need to decide who the principal star actors will be. Make sure you have some head-shot photos of them available for later reference.

If you prefer, you can write an alternative film synopsis of your own. Don't just copy a current or recent film though.

■ ACTIVITY

Read through the description of the film content text and make a list of the different colours and images that you could use as the basis of the poster. Make some notes about what the main characters might look like, and how they might be dressed. What actions might they be doing?

Marketing a movie is planned in very careful detail. First there are short news articles about a forthcoming film, and then early publicity posters and trailers. As the launch date approaches there will be adverts and interviews in magazines and newspapers, and promotional appearances by the film's stars. Just before the film opens there will be a flood of publicity gimmicks to make sure everyone is talking about and wanting to see the movie.

When planning a marketing campaign for a movie, some of the key points which are considered involve:

▷ what type of people will most want to see the film
▷ when will be the best time of year to release the film
▷ how popular it is likely to be
▷ how much to spend on promotion.

Investigation

Conduct a survey to discover more about what makes people go to see a film. Is it because of the stars, the subject matter, or because of the reviews and friends' recommendations ?

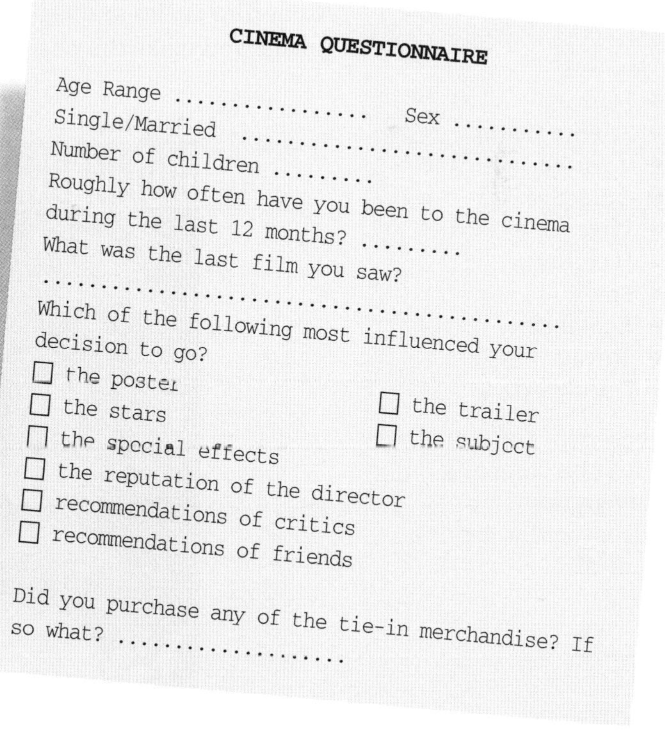

CINEMA QUESTIONNAIRE

Age Range .... Sex ....
Single/Married ....
Number of children ....
Roughly how often have you been to the cinema during the last 12 months? ....
What was the last film you saw?
....
Which of the following most influenced your decision to go?

☐ the poster
☐ the stars ☐ the trailer
☐ the special effects ☐ the subject
☐ the reputation of the director
☐ recommendations of critics
☐ recommendations of friends

Did you purchase any of the tie-in merchandise? If so what? ....

Base your survey on the questionnaire above. Try to get at least 10 people to complete it. Work in a group and combine your information. Present your findings graphically. See if you can create a spreadsheet which will enable you to do this.

Film Posters and Merchandising

Film posters exist to advertise a film. They have to be easily understood. They must attract attention, stimulate interest and suggest what type of film it is.

The poster is the central feature of any movie marketing campaign. Its image and tag line will be applied to the whole range of advertising and merchandising.

The Poster

Usually the poster will be based on a piece of original artwork or on photographs taken during production. Sometimes it will be a mixture of the two. In many cases the star actors will be featured in the poster, dressed in role and in a pose appropriate to the action or content of the plot. A strong background image will be needed to convey the atmosphere and setting of the film.

Awards and the names of the director and cast, quotes from film critics and tag lines all help reinforce the image and make the poster more effective.

Most films fit into one (or sometimes more) of the following types or 'genres':

- Horror.
- Science fiction.
- Western.
- Detective.
- War.
- Historical.
- Spy.
- Romance.
- Travel.
- Documentary.

These types of film might be done as a comedy, an adventure/thriller or a musical.

■ ACTIVITY

Think of some films on current release and make a note of which genre they would come under.

■ ACTIVITY

Look carefully at the posters on this page. Make a note of exactly what you can see in each poster.

- What do you think the story is probably about?
- What type of film is it?
- Who are the main stars in the film?
- What types of audience is the film mainly aimed at (male/female, family, age range, etc.)?
- What are the main colours used?
- What typefaces have been used?

(See pages 32 to 41 for more information about colours and typefaces.)

The Merchandise

Movie **tie-ins** are products which carry the name and motif of the film along with its characters and/or associated images. These items are commercially successful in their own right, but also provide free promotion for the film. Sometimes special joint promotion deals are made with other companies, such as fast-food outlets.

■ ACTIVITY

Extend the survey you started on page 29. Show the people you ask the posters on this page. Devise some suitable questions to discover more about the impact each poster has on them. What elements might persuade them to go to see the film?

■ ACTIVITY

Make a list of all the movie merchandising illustrated. Think of another recent film which has a lot of tie-in products and add them to your list of examples. Finally think of any examples of movie products you have at home.

Present your investigation work from this and the previous page in a highly visual way on a series of A3 sheets.

Any Colour You Like ! (1)

Colour is an essential part of the world around us. It provides us with important information about our surroundings. It affects the way we feel about and react to everyday things. Deciding on the most suitable colours for a design is a lot more complex than just picking your own favourite colour.

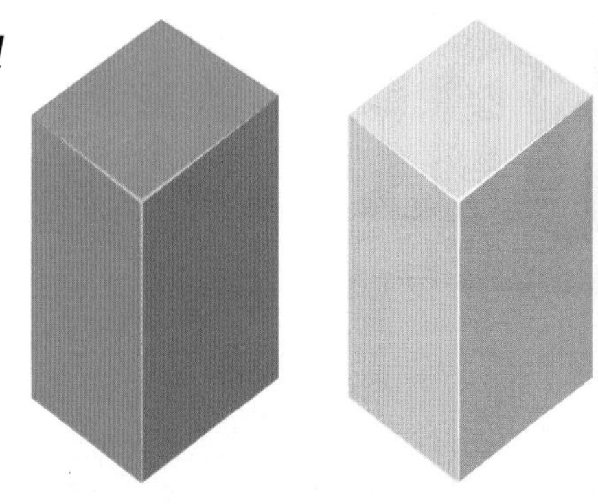

Which box looks the heaviest?

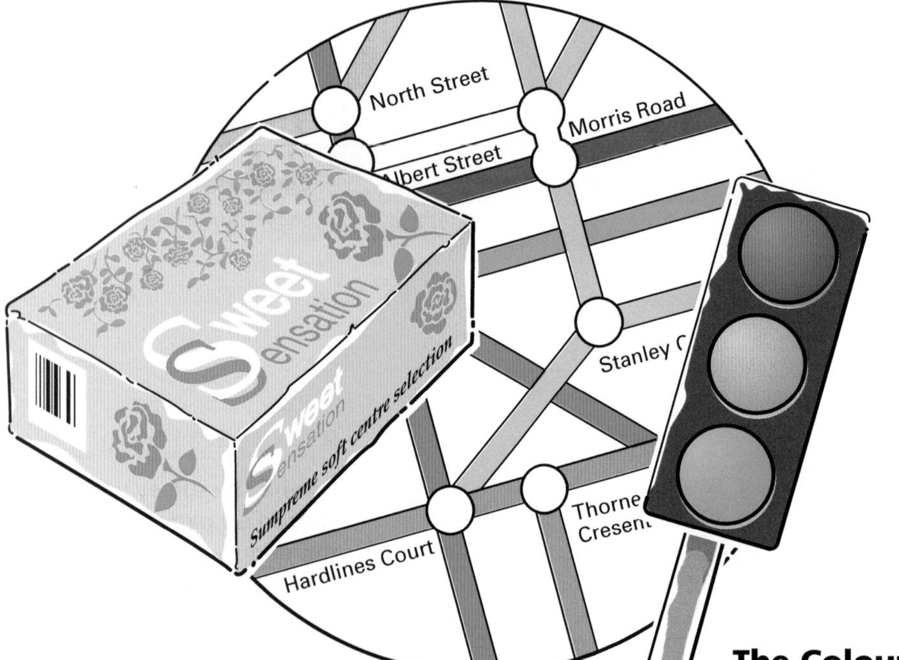

Red is the colour of fire and suggests things such as strength, love, danger and disaster and evil. In the dark, red light produces the most easily visible contrast. Red is a warm and inviting dominant colour.

The Colour Wheel

Red, blue and yellow are known as **primary colours**. They cannot be made by mixing any other colours together. **Secondary colours** are green, purple and orange, which are produced by mixing the primaries together. If you mix a primary colour with a secondary colour you will create a **tertiary colour**.

Complimentary colours are those which are opposite each other on the colour wheel, for example blue and orange. These colours create contrast, while colours which are close to each other on the wheel create harmony.

The actual colour, such as red or green, is described as the **hue**. The hue can be changed by adding **tone**, i.e. white to lighten it or black to darken it.

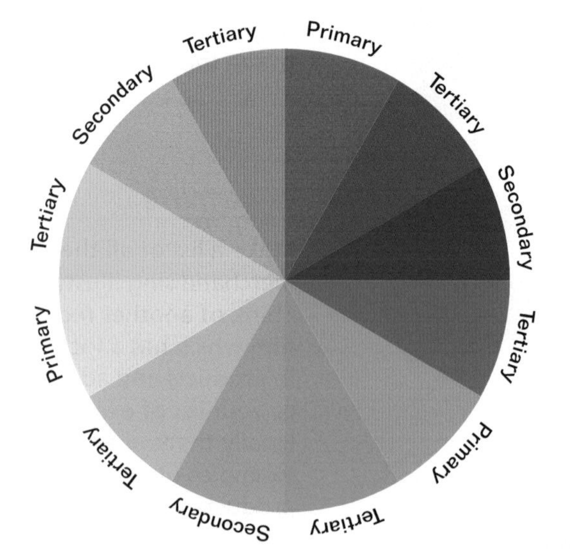

100% 0%

In a factory, productivity was increased when the washrooms were painted an unattractive shade of green, which resulted in the workforce not wanting to linger. There had also been complaints about boxes coloured dark blue being too heavy to carry. However no one seemed to notice the weight when the boxes were repainted yellow.

Designing with Colour

To create a colour scheme for a product, space or printed surface, experiment by choosing a range of harmonious hues and tones, and add a small amount of one complimentary colour. You will also need to consider the following:

- ▷ Warm colours seem to be closer to a viewer than cold colours, which seem further away. This also has the affect of making a warm-coloured object look larger than a cold-coloured object, even though they may both be the same size.
- ▷ Colours and tones can also be used to make an object look heavier or lighter.

■ ACTIVITY

Find two packages which have very different colour schemes. Identify the range of colours used, and whether each is a primary, secondary or tertiary colour. Which colours provide most contrast, and why?

Shape and Colour

Combining certain shapes and colours can produce very powerful visual images. A bright yellow bolt of lightning clearly serves as a warning. The soft shape of the cloud, combined with its warm, grey colour is much less threatening. Look for examples of how road signs and company symbols combine shape with colour.

Colour Associations

From very early times, colours have been connected in the mind with feelings, such as those of danger or happiness. Different societies and religions across the world use different colours to represent similar ideas.

■ ACTIVITY

These colour associations are not true of every culture. For Muslims, green is a holy colour, while for Hindus and Greeks white represents water. What other examples of different associations can you discover?

Orange is commonly associated with warmth and cheerfulness, sunsets and the autumn. It is a very intense colour, however, and is best used sparingly and to indicate hazards.

Yellow reminds us of the sun, and in some religions symbolises life and truth. Yellow is also the colour of cowardice, however. Yellow and black provide the greatest contrast and are therefore often used to indicate hazards. Yellow is a highly stimulating colour, so needs to be used sparingly.

Green is symbolic of the natural world. When used with blue it suggests ice, and coldness. With reds and oranges it takes on the sense of autumnal, earthy scenes. Green is used to indicate first aid and safety. It often represents growth and hope. Generally green is a calm colour.

Violet and purple are very rich colours, suggesting wealth and extravagance. They often denote royalty, and can also indicate knowledge, nostalgia and old age. Purple and violet in large amounts are unsuitable for environmental settings.

Blue is the colour of the sky and the sea, suggesting far horizons and vast depths. It often implies things like truth, peace, loyalty and wisdom. Blue is used to identify electrical equipment. Blues are easy to live with, passifying and calming colours.

Black is traditionally the colour of death and mourning in Western civilisations. Witches are black. Black and grey clothes suggest uniformity.

White is the colour of snow and the moon. It represents purity and innocence and is used for christenings and weddings. White suggests cleanliness, which is why many washing machines and detergents are white.

Any Colour You Like! (2)

Colour Fusion

When small areas of colour are seen together the colours merge and look different. For example a mixture of blue and yellow dots will give the impression of being green. This is known as **colour fusion**. Try looking closely at a large advertisement hoarding or a TV screen to see examples of this.

George Seurat used dots and small brush strokes to create his paintings. His technique is known as Pointillism.

Colour Separation

Modern colour printing uses dots of the three primary colours (yellow, magenta and cyan) to create the full range of colours when mixed together on the page. These are known as **process colours**. Colour separation is described in more detail on page 75.

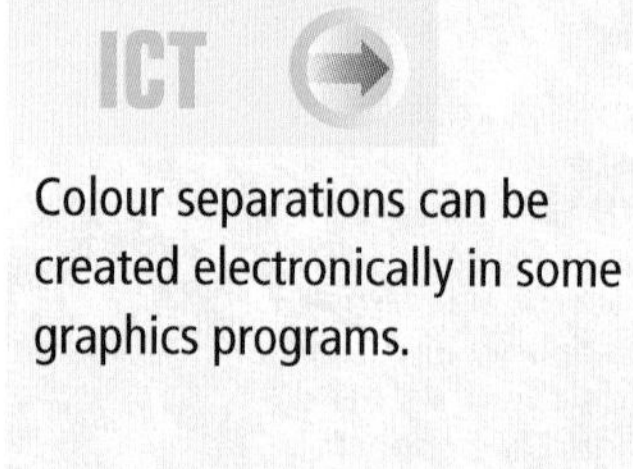

Colour separations can be created electronically in some graphics programs.

Corporate Colour

Many companies use a particular colour to help establish their brand identity. Different colours are more suited to different types of products and services.

- ▷ Red is often used to whet the appetite and to sell fast food. Pink is associated with femininity.
- ▷ Orange suggests power and is used for strong cleaners, health foods and drinks.
- ▷ Yellow represents long life (i.e. durability), sunlight, and sunsets.
- ▷ Green is the colour of nature and safety and is used to indicate 'environmentally friendly' products. It is also the colour of money.
- ▷ Blue stands for reliability, and is therefore popular with banks, travel and insurance companies, and international corporations.
- ▷ Purple expresses smoothness and excitement, and is therefore used for luxury, and exotic products.

■ ACTIVITY

Here are four products which need packaging:

- ▶ 'Zap': a household cleaner
- ▶ 'Slurp' : a strawberry-flavoured fizzy drink
- ▶ 'Rainbow': cosmetics
- ▶ 'Light and Dark': luxury chocolates.

Decide which of the containers on the right would be most suitable for each product.

Redraw them adding the name of product and background colours and graphic shapes/patterns to indicate nature of product.

■ ACTIVITY

Working as a group, make a collection of packages, logos, advertisements, etc., which predominantly use one of the colours mentioned. Create your own rainbow display.

ICT

A graphics program makes it easy to experiment with different combinations of colours.

IN YOUR PROJECT

- ▶ Experiment with different colour schemes.
- ▶ Think carefully about the things people might associate with the colours you choose.

KEY POINTS

- Colours are known as primary, secondary or tertiary.
- Colours can look warm or cold, near or far, heavy or light.
- Colours represent feelings, emotions and expectations.

Layout Grids

Behind every printed surface there is grid structure. The grid defines the height, width and position of the margins, columns and headings. Everything lines up with something.

The first stage of any graphic layout design is the development of the grid.

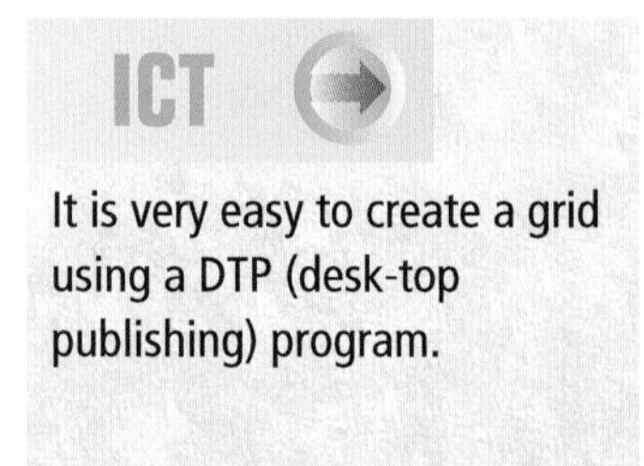

It is very easy to create a grid using a DTP (desk-top publishing) program.

Thumbnails

To create a grid you will need to explore some ideas in rough using what are known as **thumbnail sketches**. These are small, quick drawings which each represent the possible layout of the page. As you work you will need to be experimenting with the possibilities for:

- having one, two, three or more columns
- the size of the margins between the columns
- how the grid will be divided horizontally
- what graphic devices, such as lines, tints and slabs of colour, could be used
- where a running head and/or any page numbers will be placed
- the size and starting position of main headings and sub-headings
- the location of illustrations
- using a different grid for certain sections.

■ ACTIVITY

Obtain a colour magazine and place a sheet of tracing paper over a page of text.

Using a pencil and ruler, draw in the vertical and horizontal lines of the main columns of body text. Measure the width of the columns and margins.

Draw in the horizontal lines and measure the distances between them.

Repeat the process with an individual full or double-page advertisement.

See if you can work out the four-column-per-page grid design used for this book.

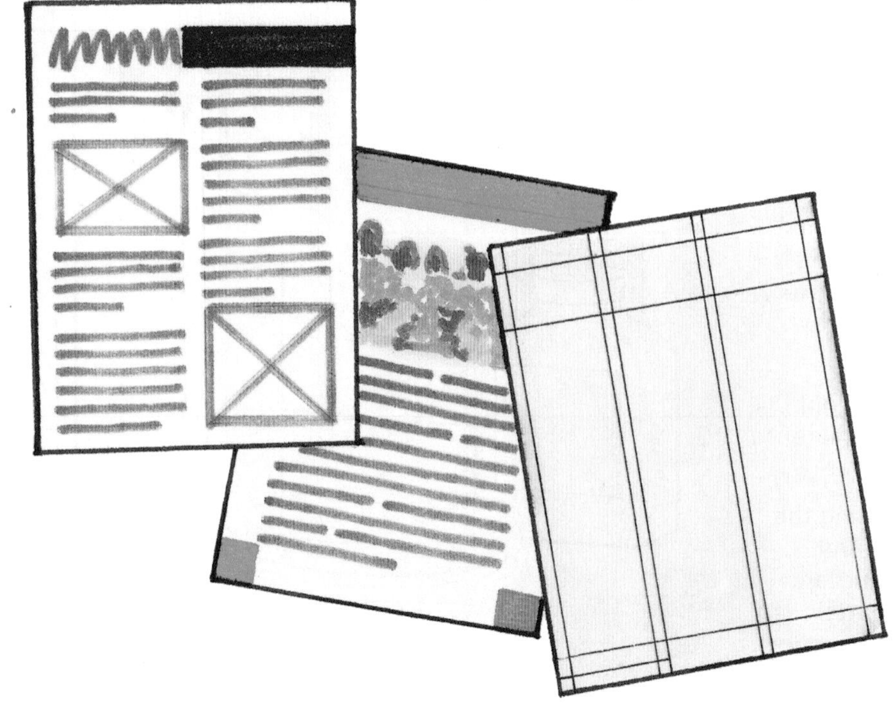

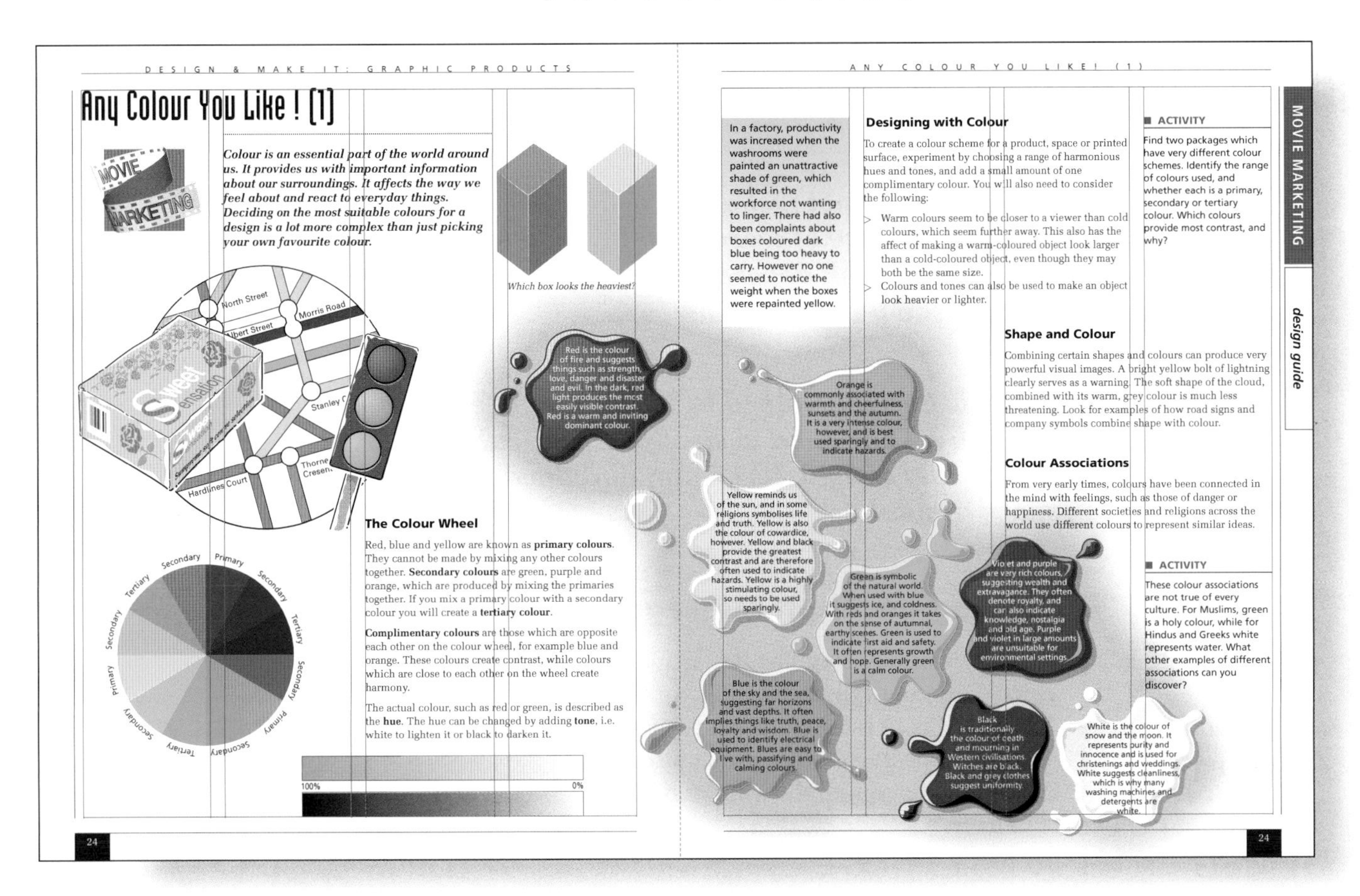
DESIGN & MAKE IT: GRAPHIC PRODUCTS

Any Colour You Like ! (1)

Colour is an essential part of the world around us. It provides us with important information about our surroundings. It affects the way we feel about and react to everyday things. Deciding on the most suitable colours for a design is a lot more complex than just picking your own favourite colour.

Which box looks the heaviest?

Red is the colour of fire and suggests things such as strength, love, danger and disaster and evil. In the dark, red light produces the most easily visible contrast. Red is a warm and inviting dominant colour.

The Colour Wheel

Red, blue and yellow are known as **primary colours**. They cannot be made by mixing any other colours together. **Secondary colours** are green, purple and orange, which are produced by mixing the primaries together. If you mix a primary colour with a secondary colour you will create a **tertiary colour**.

Complimentary colours are those which are opposite each other on the colour wheel, for example blue and orange. These colours create contrast, while colours which are close to each other on the wheel create harmony.

The actual colour, such as red or green, is described as the **hue**. The hue can be changed by adding **tone**, i.e. white to lighten it or black to darken it.

24

ANY COLOUR YOU LIKE! (1)

In a factory, productivity was increased when the washrooms were painted an unattractive shade of green, which resulted in the workforce not wanting to linger. There had also been complaints about boxes coloured dark blue being too heavy to carry. However no one seemed to notice the weight when the boxes were repainted yellow.

Designing with Colour

To create a colour scheme for a product, space or printed surface, experiment by choosing a range of harmonious hues and tones, and add a small amount of one complimentary colour. You will also need to consider the following:

- Warm colours seem to be closer to a viewer than cold colours, which seem further away. This also has the affect of making a warm-coloured object look larger than a cold-coloured object, even though they may both be the same size.
- Colours and tones can also be used to make an object look heavier or lighter.

■ ACTIVITY

Find two packages which have very different colour schemes. Identify the range of colours used, and whether each is a primary, secondary or tertiary colour. Which colours provide most contrast, and why?

Shape and Colour

Combining certain shapes and colours can produce very powerful visual images. A bright yellow bolt of lightning clearly serves as a warning. The soft shape of the cloud, combined with its warm, grey colour is much less threatening. Look for examples of how road signs and company symbols combine shape with colour.

Colour Associations

From very early times, colours have been connected in the mind with feelings, such as those of danger or happiness. Different societies and religions across the world use different colours to represent similar ideas.

Orange is commonly associated with warmth and cheerfulness, sunsets and the autumn. It is a very intense colour, however, and is best used sparingly and to indicate hazards.

Yellow reminds us of the sun, and in some religions symbolises life and truth. Yellow is also the colour of cowardice, however. Yellow and black provide the greatest contrast and are therefore often used to indicate hazards. Yellow is a highly stimulating colour, so needs to be used sparingly.

Green is symbolic of the natural world. When used with blue it suggests ice, and coldness. With reds and oranges it takes on the sense of autumnal, earthy scenes. Green is used to indicate first aid and safety. It often represents growth and hope. Generally green is a calm colour.

Violet and purple are very rich colours, suggesting wealth and extravagance. They often denote royalty, and can also indicate knowledge, nostalgia and old age. Purple and violet in large amounts are unsuitable for environmental settings.

■ ACTIVITY

These colour associations are not true of every culture. For Muslims, green is a holy colour, while for Hindus and Greeks white represents water. What other examples of different associations can you discover?

Blue is the colour of the sky and the sea, suggesting far horizons and vast depths. It often implies things like truth, peace, loyalty and wisdom. Blue is used to identify electrical equipment. Blues are easy to live with, passifying and calming colours.

Black is traditionally the colour of death and mourning in Western civilisations. Witches are black. Black and grey clothes suggest uniformity.

White is the colour of snow and the moon. It represents purity and innocence and is used for christenings and weddings. White suggests cleanliness, which is why many washing machines and detergents are white.

24

MOVIE MARKETING
design guide

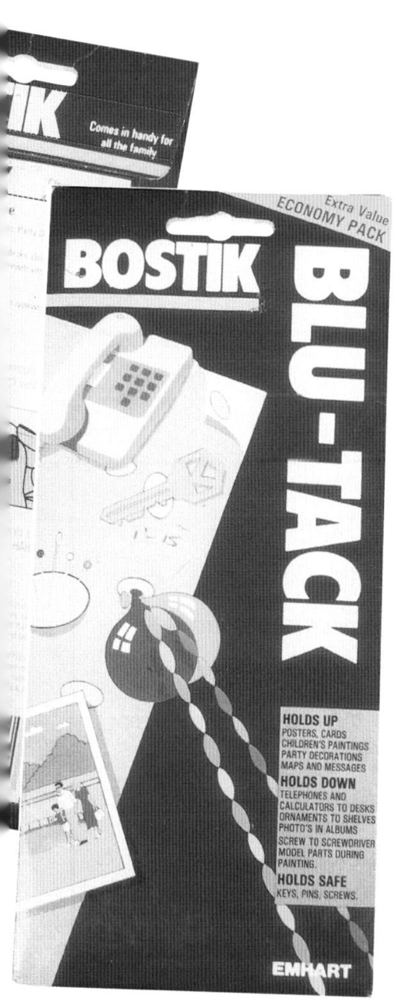

Cut and Paste

After the grid has been established, the positions of the text blocks and illustrations within the grid structure need to be explored and developed. Photographs and drawings will need to be marked up to show how they need to be scaled and cropped. Copies of the text and illustrations will then be cut up and pasted into position for checking. These sheets are often known as the **paste-up**.

IN YOUR PROJECT

- Make sure any graphic design work you do is based on a grid of some sort.
- Start with a series of quick thumbnail sketches.
- Refine your ideas using a DTP program if possible.

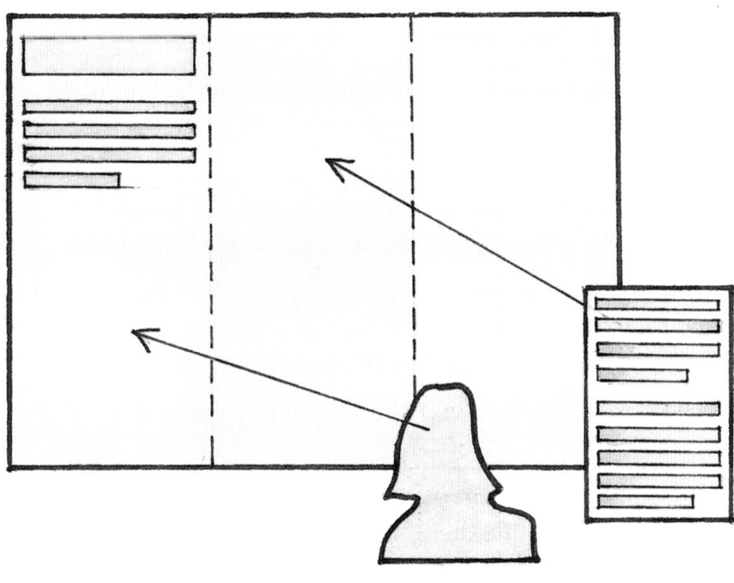

KEY POINTS

- All printed surfaces are based on a grid of some sort.
- The grid defines the position of columns of text, margins and graphic devices.
- When the grid has been finalised, the text and illustrations are positioned and checked.

Talking Typography (1)

Typography is not just about neat lettering. The appearance of words can be used to attract attention, to look decorative and interesting, and to suggest a particular time or place.

Graphic designers need to select typefaces very carefully. They need to specify a particular font, its size and weight, its colour, and the amount of space around it.

Courier can make it
Official

Times Roman speaks with
Authority

Folio Bold Condensed expresses
Doubt?

Mistral shouts
Surprise!

Kabel Medium smiles with
JOY

Futura Extra Bold Condensed screams

Sending the Right Messages

Each style of lettering has its own very distinctive tone of voice. Different typefaces can express a wide variety of feelings. It is possible to use a typestyle to help shout or whisper a message, or to attract attention to it. Alternatively you might want your words to look and sound:

▷ traditional and reassuring
▷ romantic and relaxed
▷ stimulating and exciting
▷ creative and unusual.

Meanwhile some typeface styles make us think of particular situations or periods of history.

The Wild West

Art Deco

Art Nouveau

THE 21ST CENTURY

Legibility

It is essential that typefaces are easy to read. Some typestyles are more legible than others, but their size and the space between the words and letters makes a great deal of difference.

■ ACTIVITY

Find some books which show some of the hundreds of different styles of lettering which have been professionally designed. Check what typefaces are available on the desk-top publishing programs in your school. Choose two different styles. Make an accurate copy of the word 'Typography' using each style.

Size is Important

Type is measured in **points**. One 'point' is 1/72nd of an inch (which is about 25 mm), so a 72 pt letter would be about 25 mm high.

As well as the height of each letter, the width needs to be considered. Most typefaces come in a range of widths.

The lighter or more condensed a typeface is, the larger it will need to be to be easily read. The bolder or more extended a typeface is, the smaller it can be.

The size of typeface you choose depends on:

▷ the number of letters used
▷ the amount of space available
▷ how noticeable the text needs to be.

Extra Light
Light
Medium
Light italic
Light condensed
Medium italic
Medium condensed
Medium extended
Medium Outlined
Bold
Bold Italic
Bold Condensed

Size Guide

Times New Roman (9pt)

For captions to illustrations or other short items of text an italic typeface with a size of about 8 or 9 pts should be used.

Times New Roman (11 pt)

For the main body text of a document the typeface should be between about 10 and 11 pt.

Helvetica (14 pt)

For sub-headings use a typeface which is different to the one used for the body type, at about 14 pt, possibly in bold.

Helvetica (30 pt)

For main page titles something between 24 and 36 pt is common.

Displays, signs or posters will need to use letters at least 144 pt in size.

The size guide above is only a general guide. You will need to make your own choices to suit the particular application and typeface.

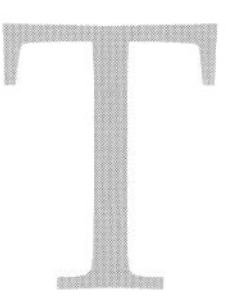

Serif – easy to read, looks traditional.

Sanserif – strong, bold and clear. Modern looking. Often used for titles and headings.

Script – looks more personal, and depending on the style used, historical. Can be difficult to read.

Decorative – attracts attention, and gives text a particular feel or association. Can be difficult to read. Best used for main titles.

Mixing It

When choosing typefaces for a publication or poster it is generally best to use no more than two typestyles. Often one will be a serif face and the other sanserif – one for body text and the other for titles. A range of sizes of the two faces can be used. These rules can be broken for special effects, providing the words can still be easily read.

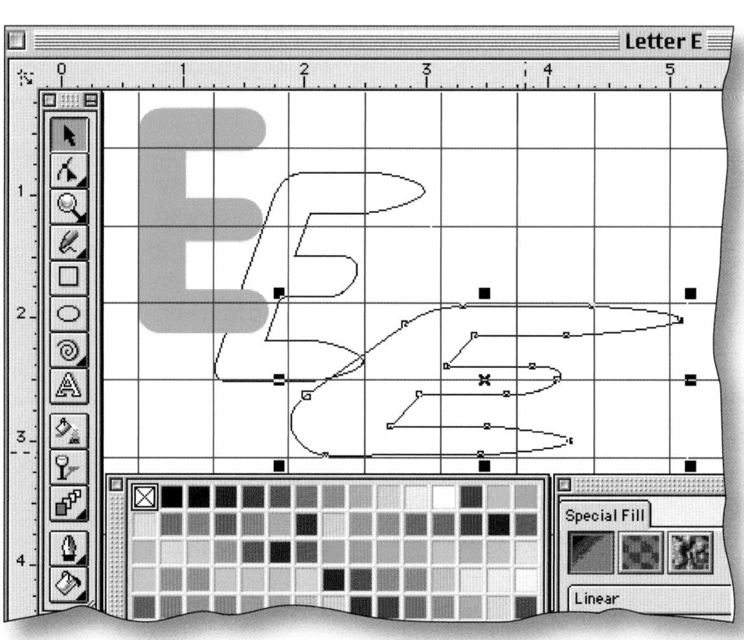

Designing Your Own Typeface

Creating a new typeface is a difficult challenge, and is best left to specialists. Most graphic designers use existing designs. One approach however can be to adapt an existing typeface, perhaps just changing one or two letters to provide something more distinctive. There are some specialist typeface design computer programs available.

Talking Typography (2)

The spaces between letters, words and lines are very important. Decreasing the amount of space between these elements can help get more text onto a page. Increasing it can help fill blank space and make the text easier to read. Both make a subtle but distinct impact on the overall design and feel of the weight of text on a page.

Places for Spaces

Experimentation with different letter, word and line spacing is needed to ensure the right effect is achieved. Word processing and desk-top publishing programs make this much easier to do.

Letter Spacing

Look carefully at the spaces between letters of the typefaces used on this page. They are close together, but don't actually touch. This is standard letter spacing. Looser spacing can be very effective for titles or sub-titles or when using condensed or decorative fonts. Very loose letter spacing can be difficult to read. Some typefaces need looser or tighter spacing for easy reading.

Loose letter spacing

Standard letter spacing

Tight letter spacing

Word Spacing

As a general guide, the ideal space between each word should be just under a half the width of the lower-case 'o'. Too little space causes the words to merge together. Too much makes a jerky read and can create distracting patterns of space, called **rivers**.

Most text is lined up to the left-hand margin.

Occasionally, for special effect, it can be aligned to the right, as these two lines have been.

When text is lined up to both sides of a column it is described as being **justified**. This produces a very strong, distinctive shape on the page, and can provide a good contrast for a particular paragraph or section. It can create problems, however, particularly if the column is narrow. Uneven and very loose spaces start to occur, making reading difficult. One solution is to hyphenate words.

Justification can create problems, however, particularly if the column is narrow. Uneven and very loose spaces start to occur, making reading difficult. One solution is to hyphenate words.

Kerning
Sometimes it is necessary to adjust the spaces between individual letters, particularly in main titles. For example the letters

V o

can be made to overlap, so as to look visually equally spaced.

Vo

In another case it may be necessary to space a series of vertical letters out more when they occur together:

Illuminate

Illuminate

The more sophisticated DTP systems will do this automatically.

Lines of text which are placed too close together can be very difficult to read and should be avoided.

Lines of text can be placed quite far apart and still be easy to read.

■ ACTIVITY

Obtain a magazine. Make a study of the letter, word and line spacing and the titles and sub-titles used throughout its content pages.

Look as some of the full-page advertisements and compare and contrast their use of the same elements.

Line Spacing

The amount of space between each line has an often surprising impact on the overall appearance and legibility of a passage of text. There are two basic guidelines:

- ▷ the line spacing should be greater than word spacing
- ▷ the longer the lines are, the more space is needed between them.

Too little line space can result in the ascenders (e.g. b, d, f, h, k, l, t) and the descenders (e.g. g, j, q, y) touching each other. Too much line spacing becomes difficult to connect together visually.

The space between lines is known as **leading** – a reference to the strips of lead which were placed between lines of metal type. Leading is measured in point sizes, like letters.

A common solution is to make the spacing about 20% more than the point size of the lettering. For example, 12 pt type would be used with 15 pt line spacing. Most DTP programs use this ratio automatically, but it is possible to change it. Using 18 pt line spacing with 12 pt lettering (i.e. 50% more) would be very readable, however. Again it's a matter of experimenting and taking into account the appearance of the particular typeface, the word spacing, the margins and other design elements on the page.

IN YOUR PROJECT

- ► Experiment with the letter, word and line spacing of your design.
- ► Think carefully about the amount of space between paragraphs and sections.

KEY POINTS

- The choice of letter styles makes a great deal of difference to the way a message is read.
- Size, styles and the space between letters, words and paragraphs all need to be carefully considered.

Paragraph and Section Spacing

If the start of a new paragraph is indented it is not necessary to leave any extra space between the lines of text. In other cases it is usual to insert one extra half measure of line spacing (e.g. 10 pt + 5 pt =15 pt). An alternative is a double line space (e.g. 10 pt + 10 pt – 20 pt). The more space which is left the greater the visual suggestion that the writer is moving on to a new topic which is not related to the last paragraph.

A new section will need extra spacing, and probably a 'sub-head' – a heading which breaks the text up and clearly identifies the start of a new passage.

Preparing Artwork

There are a variety of ways of creating type on artwork. In order of effectiveness, these are:

- ▷ Use a letter stencil (used only for architectural plans, working drawings or electronic circuits).
- ▷ Trace the letters from a typeface catalogue.
- ▷ Use dry-transfer lettering.
- ▷ Obtain a laser print-out from a DTP system.

When preparing a rough, or dummy, it is common practise to indicate areas of text by drawing lines, equal to the 'x' height of the letters, or by cutting and pasting samples of the chosen type from a magazine.

2D Computer-aided Design

To learn more about the graphic CAD programs used by professional designers visit the following software manufacturer's websites:
www.adobe.com
www.corel.com
www.macromedia.com
www.digitalvisiononline.co.uk

Computer-aided design (CAD) is a process of developing and modifying design ideas using a computer.

Electronic Imaging

An electronic image is one produced using a computer. These have revolutionised the work of the graphic designer during the last ten years. There are a number of main types of 2D CAD packages that designers regularly use.

Draw packages

Packages with the word 'draw' in their title are equivalent to using a pen on paper to 'draw' a shape and then colour it in. These also include tools to provide manipulation and tight control over lettering.

The files they produce are known as **vector** images. They store information about the length and angle of each line of an image and the colour it contains. Most clip-art is created in vector packages.

Paint or Photo Packages

Packages with the word 'Paint' or 'Photo' in their title are equivalent to using a brush on paper to create an area of colour, or to taking a photograph. The files they produce are known as **bitmap** images.

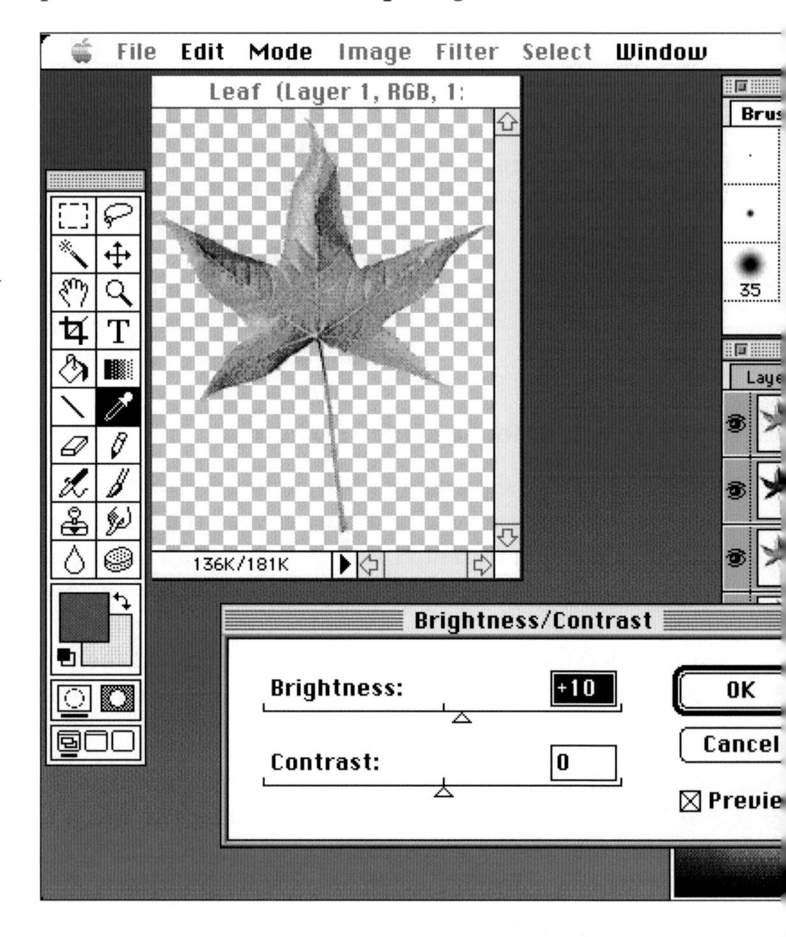

Paint or photo-manipulation programs store information by defining the position and colour of each pixel in the picture. This makes it easier to apply effects to the whole or substantial parts of the image, eg making it lighter or darker. It is possible to mask different parts of the image. The file sizes of bitmap images is much larger than vector images.

DTP packages

Desktop Publishing packages are used to assemble blocks of text and graphics together to produce a printed document. Areas of text and backgrounds can be coloured, and 'wrapped round' the outlines of images. Layout grids can be created on-screen, or pre-defined in a document template. A variety of type specifications can be pre-set in 'style-sheets', so that they can be quickly applied to a passage of text.

The text and images should be imported from word-processing or graphic files, rather than being generated within the DTP package.

The flexibility and speed of DTP systems has led to the many creative and exciting page layouts used in today's magazines, leaflets, CD covers, etc.

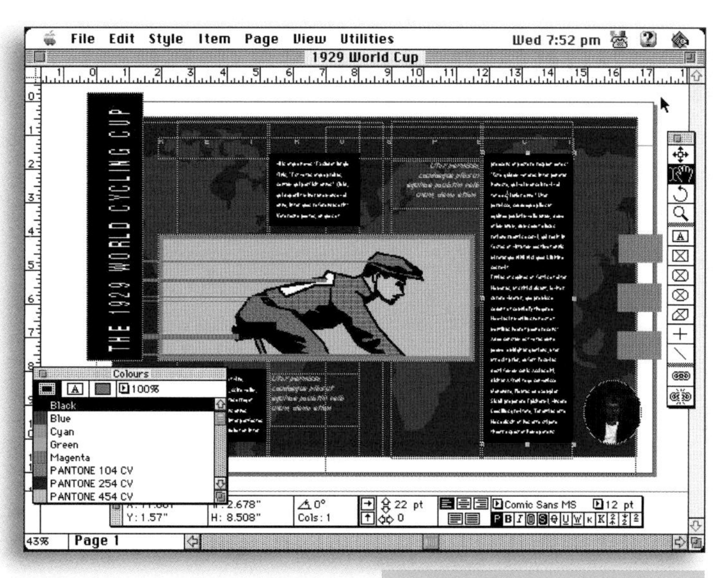

The latest versions of many of these packages include features found in other packages. For example, bitmap packages contain vector drawing tools, and can be used to create pages and animations suitable for use on the internet.

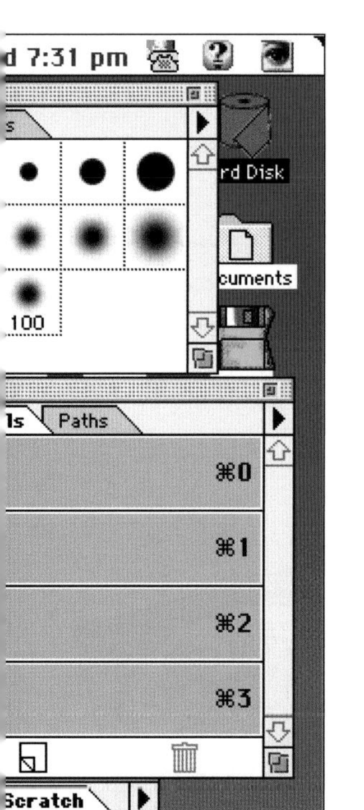

Web-design/animation

These packages are used specifically for bringing together text and graphics into 'pages' ready for the internet. The graphics can include complex animations in which the text and images can move. The text, graphics and animation are often created in other packages.

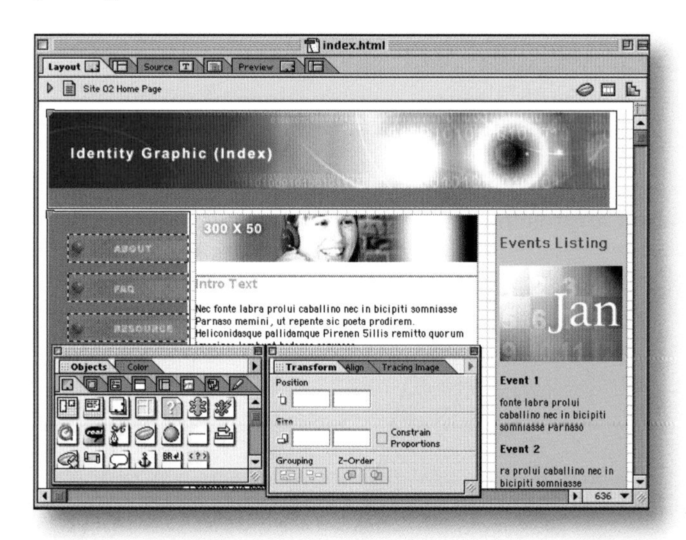

Desk-top video

These are used to edit video pictures and sounds together, and to a wide variety of special effects.

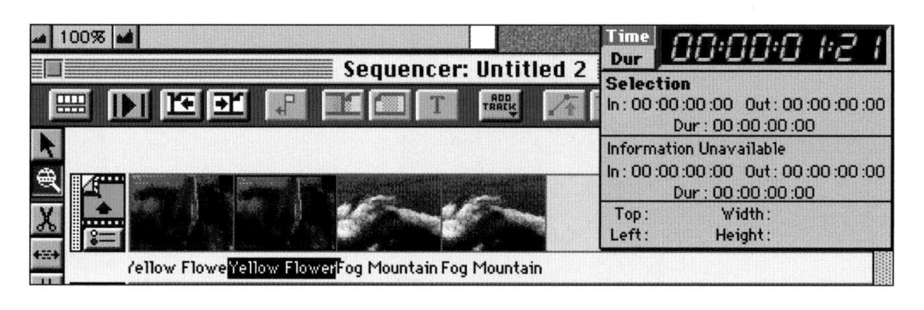

ACTIVITY

Find out what 2D CAD programs and printers are available at your school.

Choose a page from a magazine and see how closely you can copy its layout, typefaces and other design elements on computer.

IN YOUR PROJECT

- Start by sketching layout ideas on paper before using the computer
- Make sure you record the different development stages of your design. You will need to tell the story of the development of the layout in your folio

Designing and Making the Poster

Design the main publicity poster for your film. Your poster must include:

- ▷ ***the title of the film***
- ▷ ***the names of the film's main stars***
- ▷ ***the film's certificate***
- ▷ ***the tag line.***

The real size of the poster would be 1500 mm by 1000 mm (known as A4 sheet). Your artwork needs only to be approx 30 cm by 42 cm (i.e. A3).

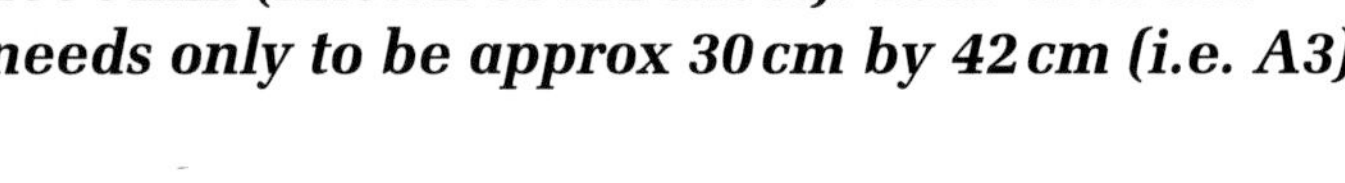

Remember to include the following things visually:
- ▶ the stars
- ▶ their roles
- ▶ the type of film

and written text to communicate:
- ▶ the film title
- ▶ the tag line
- ▶ the name of director, awards, quotes from critics, etc.

First Thoughts

Reread the description on page 30 of what makes a successful film poster.

Will the star actors will be featured on the poster? If so how will they be dressed and what will they be doing? Where is the film set?

A strong image will be needed to convey the atmosphere and setting of the film. You will need to develop something that expresses what type of movie it is and conveys its atmosphere and setting.

Experiment with different combinations of colour, texture and tone.

Make sure you come up with ideas for at least three different approaches for the poster design.

Developing the Details

When you have decided which approach to follow, work out the dimensions of the underlying grid (see page 36). Experiment with and choose a range of typestyles and weights for the main title and other texts.

Draw out and colour in a rough, but neat, A3 or A2 version of the poster and ask a number of people to evaluate how well it works. Make any necessary changes.

The Final Artwork

Think carefully about the media you will use (see pages 112 to 115) and the paper you will work on.

Produce the final version of your poster to the highest possible quality. You might be able to use a computer graphics program. If possible, arrange to make a colour A4 or A3 photocopy of your final artwork.

You could also design a similar but different poster for use in magazines or other possible locations which would require alternative sizes or formats – the front of a bus, side of a taxi, etc.

Graphic Identity (1)

Most organisation use a visual symbol or logotype to enable the public to rapidly identify the name of the company. The best symbols are simple and distinctive, and also manage to say something about the nature and quality of the products or services on offer.

Design Approaches

Most graphic identities are based on one or a combination of the following:

- ▷ a design which uses the initial letters of the name of the organisation (known as a logogram)
- ▷ a design in which the name of the company is written out in full in a specially designed and highly distinctive typeface (known as a logotype)
- ▷ a design which is a simplified illustration of the product or service the company offers (known as a symbol)
- ▷ a design based on a decorative shape, pattern or form.

Colour also plays an important part in the design of visual symbols. (See pages 32 to 35).

COUNTRYSIDE
COMMISSION

SKY

Bhs

■ ACTIVITY

Look carefully at the identities on the right. What approaches to design are they each based on?

Make your own collection of symbols and logos. Say how effective you think each is.

- ▶ Is it easy to identify?
- ▶ What does it tell you about the company?
- ▶ Why have its colours been chosen?
- ▶ How would you describe any typefaces used?

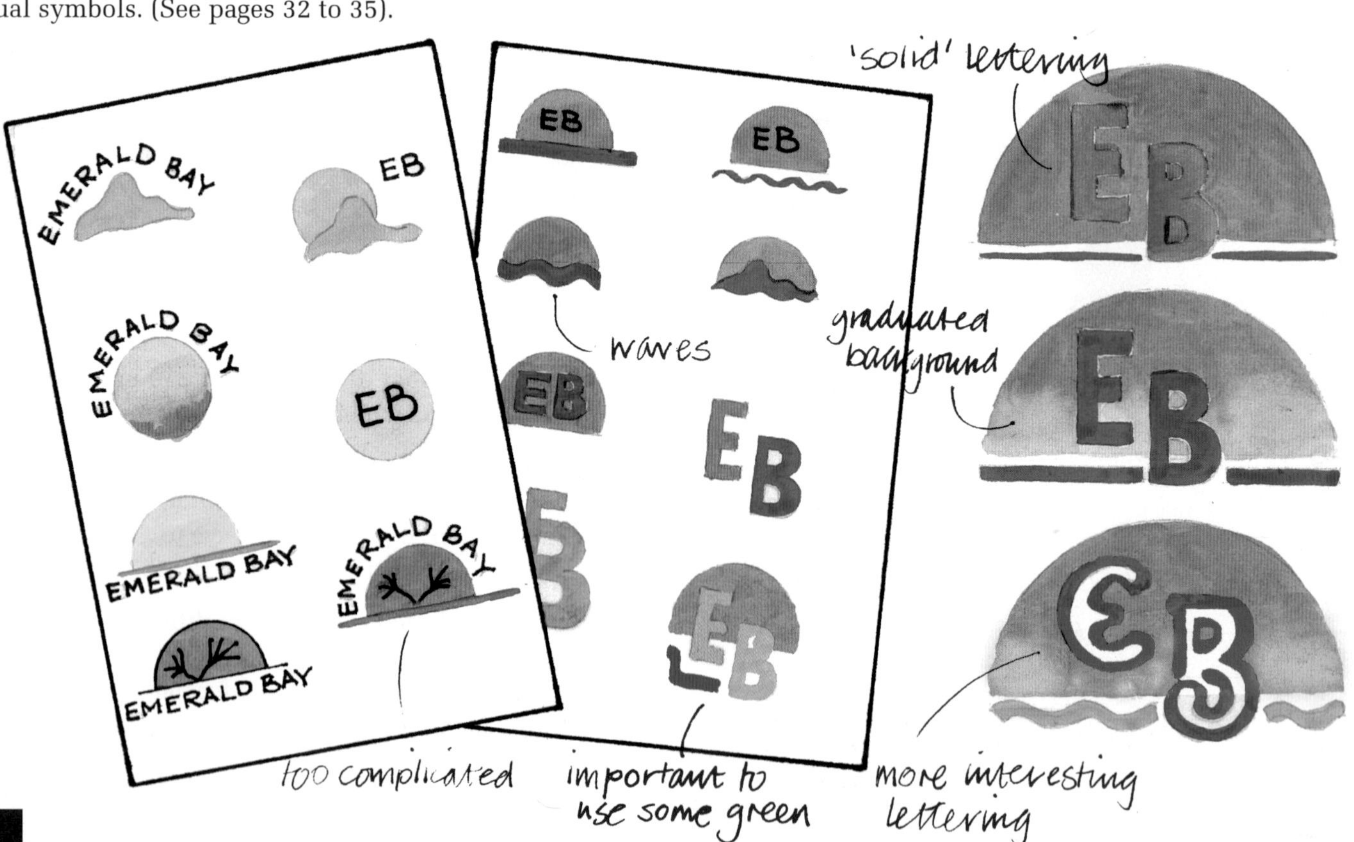

■ ACTIVITY

Study the letter heads and business cards. Try to obtain some other examples of your own. Describe and evaluate their:

- layout
- choice of typeface and style
- use of graphic devices (e.g. lines, textured areas)
- use of colour
- the design of any logo or symbol which has been used.

■ ACTIVITY

Design some form of full-colour symbol or logo based on your home address. You could experiment with ideas derived from:

- the number of your house
- the name of your house
- the appearance of your house
- a decorative detail of your house
- a family pet
- an object or event which represents the whole family.

You should also consider including simple graphic devices such as lines, squares, circles, etc.

When you have finalised the design, produce neat versions of it to show it in use as a letter-head on plain white A4 paper, and as a personalised business card, 90mm by 50mm. All lettering will need to specified as an exact typestyle and size, and should preferably be done by computer.

Finally, take your designs to a local copy shop and ask for quotations for short and long runs for black and white and colour versions, and different qualities of paper and card. Present this information in a table, and discuss your findings with your family. Record their conclusions about your design and the most appropriate print run and type.

Graphic Identity (2)

Organisations employ designers to create new corporate indentities to help them launch or revitalise their business and create more interest in their products. Corporate identities are a form of self promotion that works continually for the company.

Corporate Identity

Corporate identity is the application of a logo or symbol across a range of applications such as letter heads, business cards, shop signs, vehicles, clothing, etc. The detailing of the designs often needs to be modified for each application. These are documented in a special manual to show exactly how the various aspects of the design should be used in different situations.

Some of the best designs are those that are flexible enough to evolve to suit a set of changing circumstances, for example when a new product is added to the company's range.

Making a Name for Yourself

The easy part of brand naming is finding something short, easy to pronounce and memorable. The hard and expensive part is avoiding unfortunate different meanings of the name in different languages, and making sure it can be registered as a trademark.

Getting the right associations is the next step. Sainsbury's soft drink Gio has a get up and go, holiday, beachy feel to it. It called its cola Classic because it wanted to send the message that it is as good as anything else on the market.

But Persil and Ariel tell us nothing about soaps. We accept without question that Typhoo is a tea, Anchor a butter, Apple a computer, Orange a mobile phone. If someone offered you a food product called Frog's Nose you would probably be disgusted – but you don't worry about eating Bird's Eye foods!

Alan Mitchell

Product and Brand Names

Names are important. They say something about the nature and quality of a product or company. Brand names identify the company, while product names identify a specific model or type of product they make.

One approach to coming up with a name for something you have designed is to make a list of words which describe what the product is. Then make another list of words which describe what the product looks, feels, tastes, smells and/or sounds like. Try combining different words from your lists together. Words which sound similar and begin with the same letters often work well together.

Designing and Making the Movie Merchandise

Some suggestions for merchandising:

- ▶ watches
- ▶ fun slippers
- ▶ plastic figures
- ▶ flasks
- ▶ party crackers
- ▶ paper napkins
- ▶ balloons
- ▶ sweets
- ▶ pin-ball machines
- ▶ socks
- ▶ hair-care bags
- ▶ imitation credit cards
- ▶ stickers
- ▶ greetings cards
- ▶ toothbrushes
- ▶ key-rings.

First Thoughts

What ideas have you got for possible tie-in products? Make sure your ideas are appropriate for the type of film and the sort of audience who will see it.

You will need to design a special motif (a visual identification symbol) to use on the various products (see page 46).

Sketch and develop arange of suitable ideas.

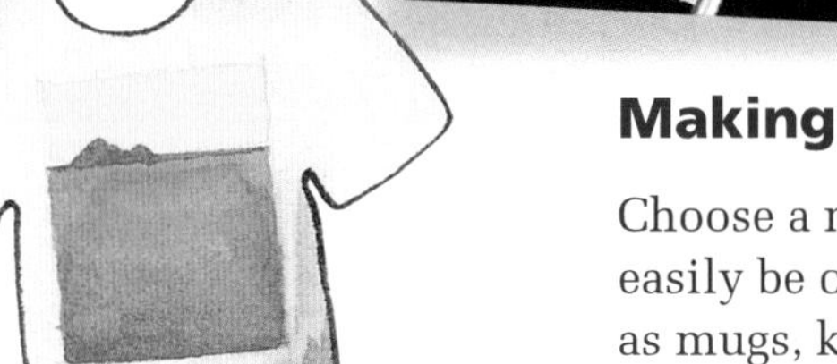

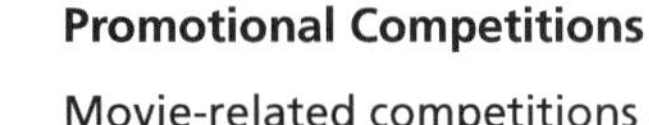

Making It!

Choose a number of ideas to make up. Artwork can easily be cut out and applied to existing products such as mugs, key-rings, paper plates, etc. You may be able to create the motif in fabric and sew it onto a bag, T-shirt or other garment. If you have time you could go on to design and make some of the items suggested below.

Promotional Competitions

Movie-related competitions are sometimes devised in association with magazines, radio and TV shows and product manufacturers. The questions and prizes relate in some way to the film and its stars.

- ▶ What suggestions have you got for a suitable competition?
- ▶ What do people have to do to win?
- ▶ What is the prize?
- ▶ Where will the competition be placed?

Ideas for a Press Pack

Press packs are sent to journalists to give them information about a forthcoming film. This material often includes:

- ▶ star interviews
- ▶ trailer
- ▶ press advertising
- ▶ TV advertisements
- ▶ radio advertisements
- ▶ merchandising details
- ▶ dates and locations of promotional events
- ▶ competitions.

Journalists are free to use these materials in articles or radio and TV shows.

Devise and make a press pack to promote your film.

Teaser Trailers

A teaser trailer is released several months before a movie's release, often to co-incide with a current major movie. It helps establish the film as a forthcoming event. The teaser should not give too much away – just enough to suggest what's coming. A 30 second sequence needs to contain a good piece of music and some strong, memorable images combined with a hint of adventure. Introducing the tag line is also important.

Prepare an outline storyboard for a teaser trailer for your film.

Designing Point-of-Sale Displays (1)

New products are often promoted using a point-of-sale display structure which will catch the shopper's eye from a distance. Such displays need to be simple and cheap to make as they will not be used for long. Bold, bright shapes and colours are needed to make them stand out.

DESIGN CHECKLIST

The main considerations for developing point-of-sale displays are:

- ✓ the type of product to be sold
- ✓ who the product is aimed at
- ✓ the competition
- ✓ the situation of the selling place
- ✓ the maximum number of products it must be able to hold at any one time
- ✓ the budget available
- ✓ the space available
- ✓ how the display will be made
- ✓ how many will be made.

Design Requirements

The term 'point-of-sale' refers, quite literally, to the place in which a customer will decide to purchase the product – not just the shop, but the specific areas in which products are displayed and where money and goods are exchanged – often a counter, or sales desk.

Newly launched products, or those with a special promotional or sale-price offers, are promoted by means of a special display. The displays are only likely to be needed for a limited period (e.g. 3 weeks) in all a retail chain's stores (which might typically range from half-a-dozen shops to several hundreds).

A point-of-sale display might simply be an arrangement of the products themselves, possibly in a specially designed container or dispenser. It might also involve a poster, a three-dimensional item, a cardboard cut-out, stickers on the window, or special signs. The choice is endless and only restricted by imagination and cost.

Often these displays are interesting shapes involving complex nets, folds and cut-outs. The most important feature of a point-of-sale display is that it is noticed and seen by potential customers, so correct positioning is essential. It could be at a suitable height at the entrance to a shop, or on the forecourt of a garage, or wherever people are most likely to pass or linger. To help attract attention it can be brightly lit, or even musical!

■ **ACTIVITY**

Make a study of a point-of-sale display. You may need to visit some local shops and ask permission to examine one of the displays. Look out for units which promote new books, CDs, videos, cosmetics, special offers, etc.

Prepare a coloured drawing of the unit on A3 paper, labelled with your comments about how you think it has been made and how effective it is. Base your text on the information about designing point-of-sale displays on this page.

Production Requirements

The batch production of point-of-sale displays is an essential consideration. The numbers required will be limited, and therefore they will be relatively expensive to produce. Some may even be 'one-offs', perhaps made by hand. Some may need to be made from card, or use more expensive materials, such as acrylic sheet, wood, or metal. They may have to last a long time or be reused, or they may only be required for a day. The designer needs a clear statement of the number required and their expected life-span.

Suitable printing processes need to be identified. Screen printing is suitable for short production runs and for printing a range of different materials, and produces bright and bold shapes and colours very effectively. (See page 81.)

Point-of-sale displays have to survive in quite hostile situations, they have to resist the everyday rigours of life in a shop, bright sunlight in the window, dust, etc. They have to be easy to install, keep clean and restock as appropriate.

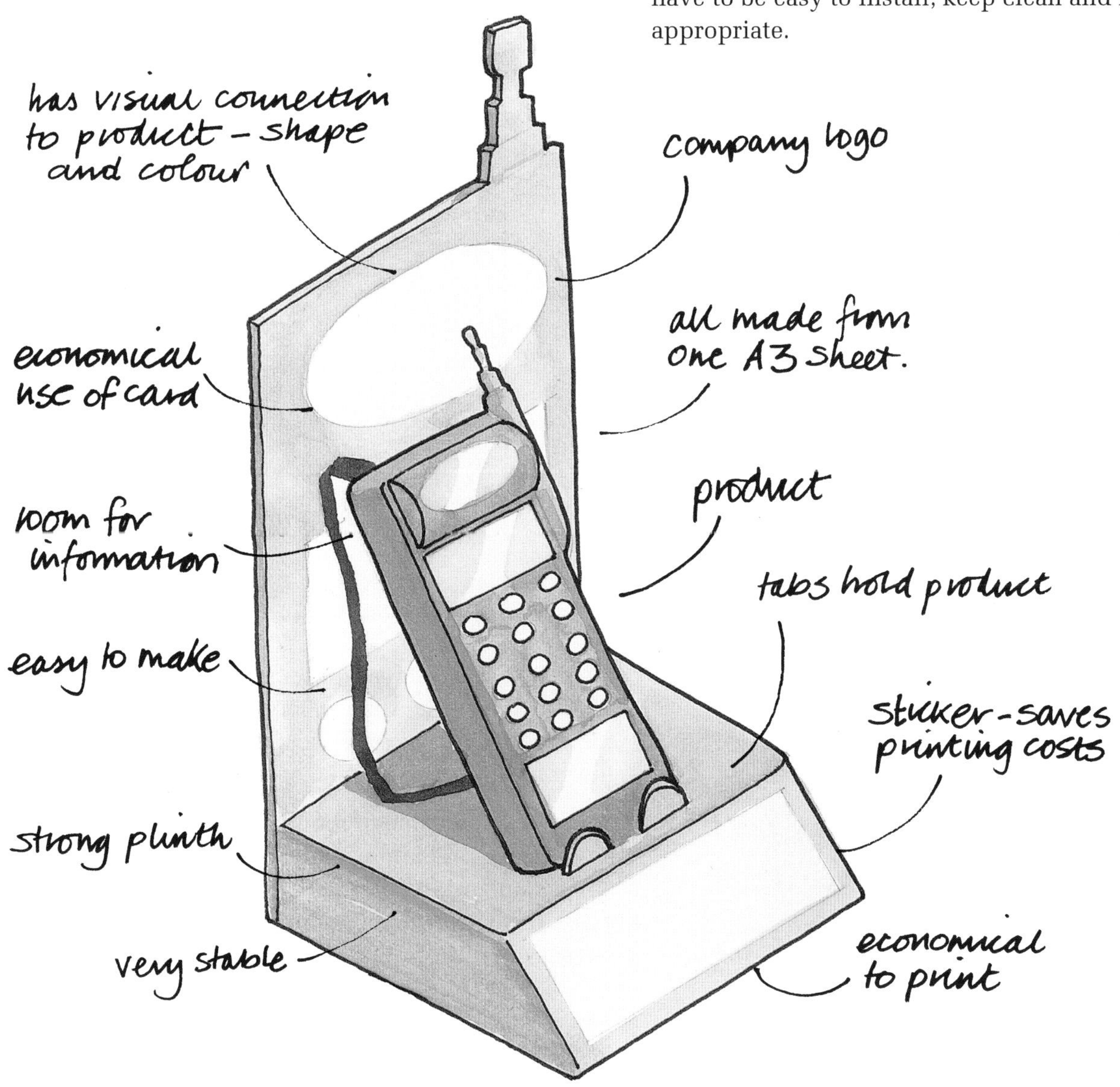

Designing Point-of-Sale Displays (2)

Design a point-of-sale display unit to promote either the launch of your film or of the video release of your film. The unit you design will be placed inside the cinema foyer.

Designing and Making Your Point-of-Sale Display

1 Begin by designing the basic net shape. You could copy the one below out onto thin card (or use a photocopier), and make it up as shown.

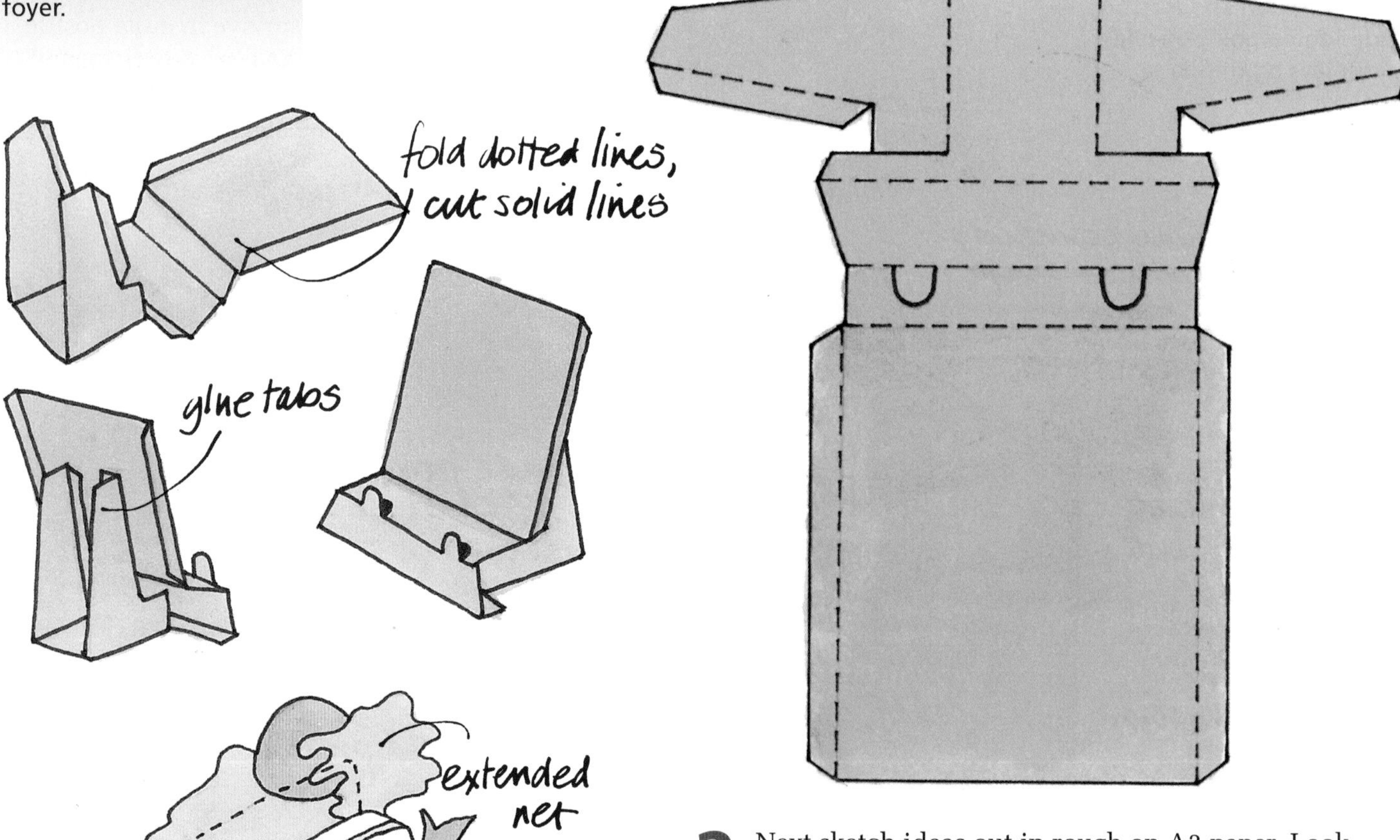

2 Next sketch ideas out in rough on A3 paper. Look closely at the shapes, forms, colours and textures of the product you have chosen. Use these as the basis for the graphic design of the display.

Make sure the name of the film is highly visible. Include any other written information you think is important. Use a standard net design to base your ideas on, extending and reshaping the back plane and/or cutting shapes out.

As you develop your ideas, consult the checklist on page 50 to check you are considering all the necessary aspects of the design. Make sure you think about the cost of materials, printing processes and other manufacturing costs.

3 Make a new version of your net in paper. Before you cut it out and fold it up make any necessary alterations to its shape. In rough, draw out the graphic design.

Then test it by making it up. Make a careful note of any modifications you wish to make.

ICT ➔

Nets can be very effectively designed and printed out using ICT.

4 Depending on the size of the product you choose you may decide to make a final full-size display or a scale model.

Prepare a new net in card. Draw on and colour in the graphics in full, paying particular attention to accuracy and neatness. Do not cut or fold the net until the graphics have been completed.

Cut the finished net out very carefully and precisely. Score folds and creases, and try and achieve neat, clean folds.

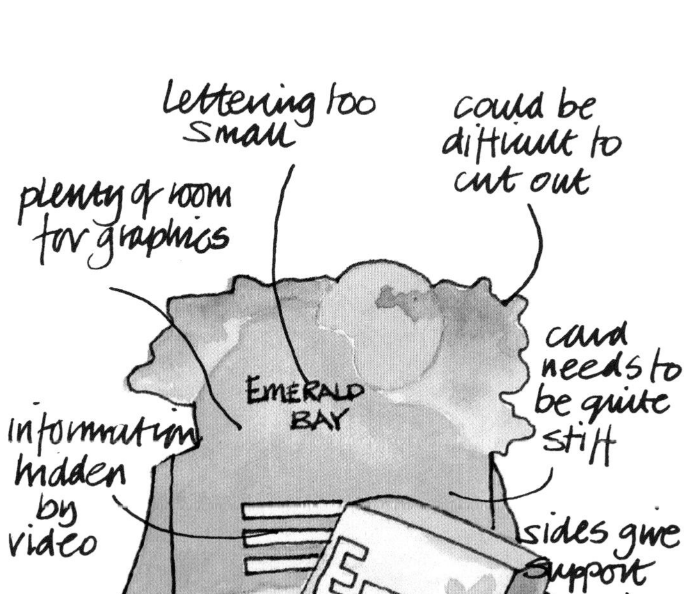

Testing and Evaluation

▷ How long does it take to assemble the unit?

Place the product, or products, in the display unit.

▷ Does it hold them securely?

Take the unit along to an appropriate location and ask if you can place it on the counter, or at some other suitable place for a short period of time.

▷ How well can it be seen?
▷ Does it attract the attention of passers-by?
▷ Does it appear to be strong and durable enough?

If possible, obtain and record the opinions of the shop manager and some customers.

Take photographs of the unit in position.

Examination Questions 1

Before spending about one and a half hours answering the following questions you will need to do some preliminary research into charities.

To complete the paper you will need some plain A4 and A3 paper, basic drawing equipment, and colouring materials. You are reminded of the need for good English and clear presentation in your answers!

Promoting a charity can take a variety of different forms: posters, collecting boxes, TV advertisements, items for sale to the public, etc.

Most charities have distinctive logo or graphic image which they use on their products and promotional lines. These help make the charity easy to recognise.

Choose a number of charities and make a study of:

- the many different graphic images used as part of a corporate identity to promote the charity
- the variety of products and promotional items sold to generate income for charities.

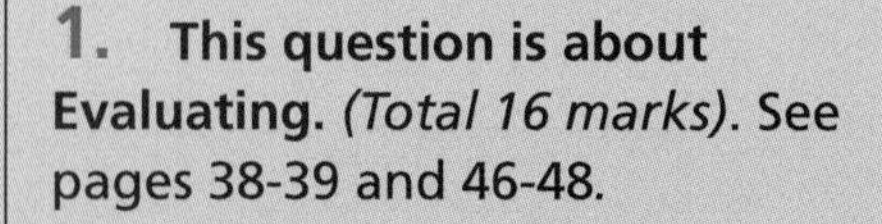

1. This question is about Evaluating. *(Total 16 marks)*. See pages 38-39 and 46-48.

Study the charity logo/graphic image below and answer the questions that follow.

(a) Most organisations use a visual symbol or logo. Why is this? *(2 marks)*

(b) How does the image show that it is related to promoting good health? *(2 marks)*

(c) Name another feature that could be used to form a logo promoting a medical charity. Make a simplified drawing of this feature. *(4 marks)*

(d) Colour is important in graphic design. Name the three primary colours. *(3 marks)*

(e) Name the colours associated with signs for:
(i) a first aid kit
(ii) a hazard warning sign
(ii) a 'take-away' food shop.

Give a clear reason for each of your choices. *(3 marks)*

(f) Name the colours:
(i) considered holy to Muslims,
(ii) that represents purity and innocence to Westerners and water to Hindus. *(2 marks)*

2. This question is about Evaluation. *(Total 10 marks).* See pages 50-53.

A charity is running a fundraising activity in which the public are asked to donate unwanted audio tape cassettes.

It has special collection units to be placed in supermarkets. Study the drawing below.

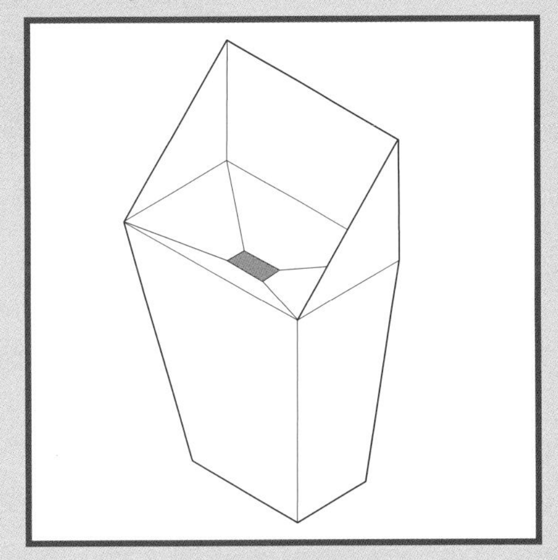

(a) What features help make the collecting unit easy to recognise? *(2 marks)*

(b) What could be improved about identifying where to put the cassettes? *(2 marks)*

(c) Describe and explain two other possible design faults with the unit. Briefly suggest how the design could be improved. *(6 marks)*

3. This question is about Designing and Market Influences *(Total 34 marks).* See pages 32-43, 46-48.

An environmental charity called 'Save Our Forests' is concerned with protecting and managing the Earth's forests. The charity has asked you to design a logo as part of it corporate image. You also need to show how the logo will be used on its stationery, transport and promotional items.

Save Our Forests has provided the following specification to help you:

- the initial letters (SOF) must be included in a suitable typeface (font)
- a logo or pictogram must be included. This must reflect the purpose of the charity
- no more than three colours are to be used.

On A3 paper use annotated sketches to:

(a) Provide three different initial ideas for both the logo and the letter style for the logo. *(14 marks)*

(b) Select the best elements from your initial ideas and produce a final, coloured presentation drawing of your logo for the charity. *(6 marks)*

(c) The charity's workers need an identification card (85x55mm). Design an identity card using your own name and your logo design. It must have a space for a photograph (25x25mm). Apply your colour scheme. *(6 marks)*

(d) Redraw or trace the outline of the charity's vehicle (shown on the right) and show how it would look in 'Save Our Forests' corporate style. *(5 marks)*

(e) The charity would like to have the words 'Save Our Forests' added to their vehicle. Briefly describe how ICT could be used to produce this sign. *(3 marks)*

4. This question is about Evaluation. *(Total 9 marks).* See pages 24-25.

(a) How could a questionnaire contribute to an evaluation of your ideas for a logo? *(2 marks)*

Compare your final design with the original specification and then complete the following sentences:

(b) My graphic design is successful because… . *(2 marks)*

(c) My lettering is successful because… . *(2 marks)*

(d) If I were to change anything about my identity card it would be… , because…? *(3 marks)*

5. This question is about Processes and Manufacture. *(Total 16 marks).* See pages 42-43.

(a) What do the letters DTP stand for? *(3 marks)*

(b) Give three ways by which DTP could help you prepare a newsletter for the charity *(6 marks)*

(c) How can photographs be put into a charity's newsletter? *(3 marks)*

(d) How can the completed computer designed newsletter be transferred electronically to the printers for processing? What is the advantage of this over a paper copy? *(4 marks)*

Total marks = 85

Project Two: Introduction

NOVELTY PACKAGING

You work for a design consultancy group which specialises in packaging. Carefully read the following letter which you have received from Paper Chain, a chain of high-street stationery and gift stores.

Existing Solutions (page 58)

Quality Counts (page 86)

Production Planning Systems (page 81)

Paper Chain Stores
London

Further to our recent meeting I write to confirm that we would like you to develop design ideas for a range of novelty-item packages to be sold as potential gifts in our shops. The idea is that each pack will contain two items which are related together in some way through a familiar name or phrase. This should be visually emphasised through the package graphics. Please note we are particularly concerned about environmental issues.

- The package should be unusual in appearance. This should be achieved either by using a bubble-pack or a non-standard net which enables part of the products to be seen on display.
- You will also need to design the surface graphics for the package. This must include its name and all other necessary information about the products it contains.
- An outline of the printing and package production processes will also be needed, together with the cost implications for the potential lengths of production run.

You will need to provide us with:

- a full-scale package mock-up, using a computer-graphic system where possible
- a short report providing full details of the printing and packaging processes and costs involved.

At this stage we would envisage a first production run of 1000 units, but would welcome your advice on the cost implications for runs of 500 and 5000 units.

We very much look forward to seeing your ideas.

UNDERSTANDING THE BRIEF

- ▶ What is the key idea of the novelty packaging?
- ▶ What will help make the package eye catching?
- ▶ What graphics and text will be on the packaging?
- ▶ What manufacturing and costing details do you need to provide?

First thoughts

- ▶ What two items could be suitably packaged together?
- ▶ Make a list of phrases and sayings.

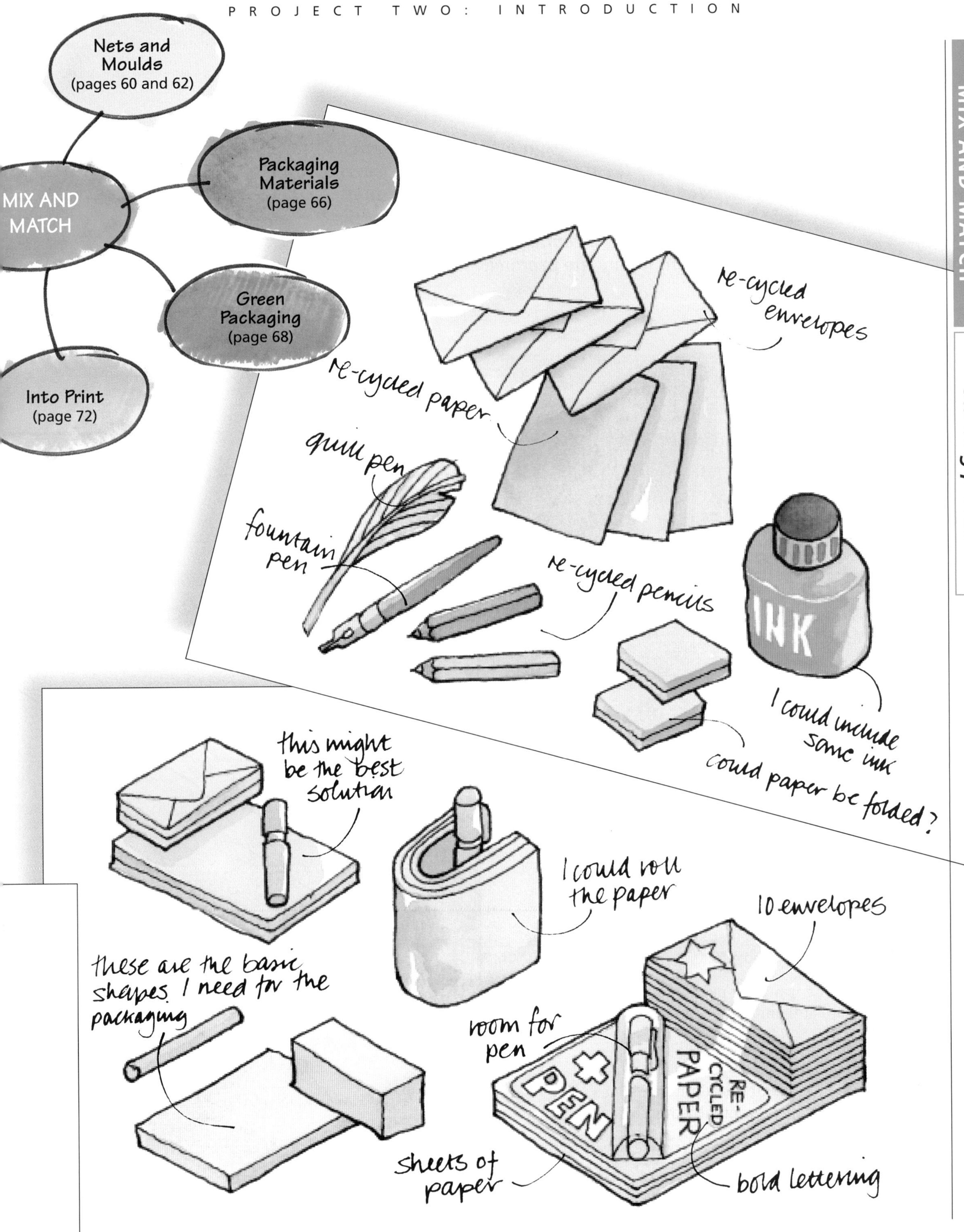
Nets and Moulds (pages 60 and 62)
MIX AND MATCH
Packaging Materials (page 66)
Green Packaging (page 68)
Into Print (page 72)
re-cycled envelopes
re-cycled paper
quill pen
fountain pen
re-cycled pencils
INK
I could include some ink
could paper be folded?
this might be the best solution
I could roll the paper
10 envelopes
these are the basic shapes I need for the packaging
room for pen
PEN
RE-CYCLED PAPER
sheets of paper
bold lettering

On the Shop Floor

Packaging serves to protect its contents, to contain them so they can be easily carried, and to identify the contents of the container. Packaging also serves to help attract the consumer's attention and help persuade them to buy the product.

To learn more about the packaging industry, visit **www.incpen.co.uk**

Find out about barcodes and how they work at: **www.beakman.com/upc/barcode.html**

■ ACTIVITY

Visit a local stationery or gift shop. Ask permission to undertake the following study activities.

▷ What range of items does the shop sell?
▷ Make some sketches of examples of the different ways in which items have been packaged.
▷ How are the products displayed – piled on shelves, in special racks, in boxes?

Spend some time watching people look around the shop (try not to stare at them!). Make notes of:

▷ which displays attract their attention
▷ which items they pick up and look at
▷ how long they spend examining the items
▷ which, if any, they eventually buy.

Remember to present all your work using neat, coloured sketch drawings on A3 paper.

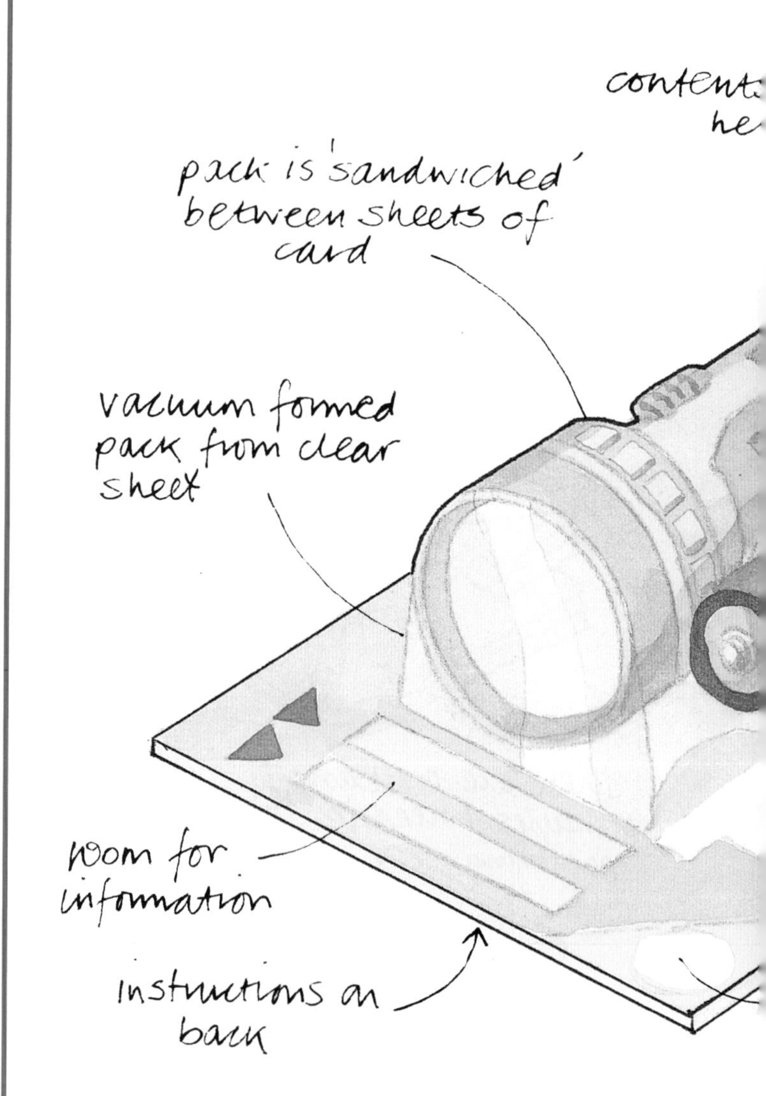

- ▶ How well does the packaging protect its contents?
- ▶ How could the package be displayed in the shop?
- ▶ What are all the different materials the packaging has been made from?

- ▶ What shape would the materials have been before they were manufactured?
- ▶ How might the package have been printed?
- ▶ Have any special effects been used?
- ▶ How are you told what the product in the packaging is?

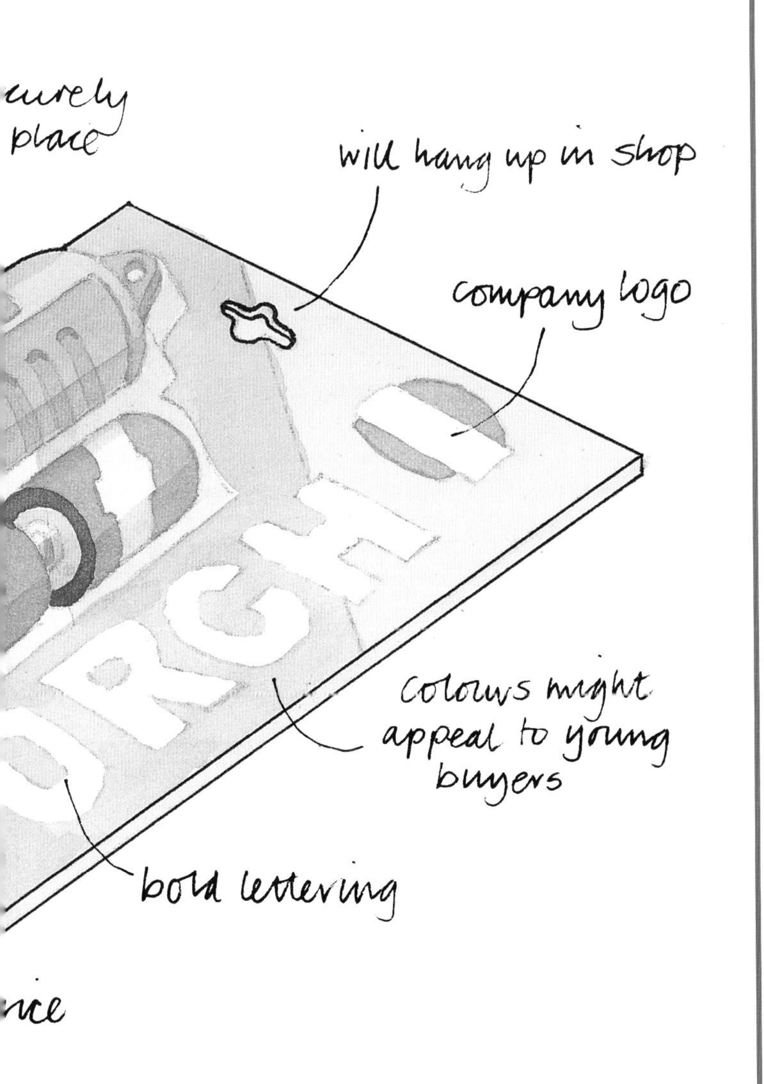

Investigation

You will need to find out more about how things are packaged and displayed in shops, and the sort of things stationery shops sell.

Packaging Disassembly and Evaluation Activity

Obtain an item of packaging which is no longer needed. It should be something more unusual than a conventional six-sided box. Try and find things such as gift boxes with unusual lids or methods of closing, more expensive chocolate or Easter-egg boxes, or packages where the product can be seen.

Compare the design features of your package with some of your friends'. How have different designers and manufacturers solved similar problems?

Present your study by means of a coloured A3 drawing.

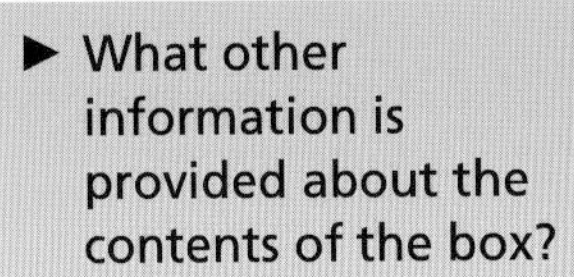

- ▶ What other information is provided about the contents of the box?
- ▶ What is the name of the manufacturer, and the name of the product?
- ▶ What necessary legal information is included?
- ▶ How might ICT have been used to design and make the package?
- ▶ How would you describe the style of lettering used to write the name of the product?

- ▶ What moods and feelings do the colours, images and lettering suggest?
- ▶ Describe the sort of person who might buy the item, and for who, and on what sorts of occasion?
- ▶ Overall, do you think the packaging is a successful piece of design?
- ▶ What are its good and bad points?
- ▶ Can you suggest any improvements?

Designing and Making Nets

A net is the flat shape of card, paper, plastic or other material that is designed to be folded up into cartons, boxes, packages, display stands, etc. It is sometimes called a development. The net is printed and then pre-cut and scored, ready to be folded up.

A lot of creative design work goes into creating the correct shapes and images.

Nets can be very effectively designed and printed out using ICT. More advanced packages will show what a net would look like when folded up, and will let you add a graphic image to its surface.

■ ACTIVITY

Obtain a package which is no longer needed. Try to find something which is more than just a regular six-sided box. Without cutting it, carefully unfold it back into its original flat form.

Study the net carefully. Look closely at the way in which surfaces have been designed to fold and slot together, or have been stuck for extra strength.

▷ How large are any flaps?
▷ How thick is the card?
▷ How have unusual shapes been created through the use of irregular cuts and folds?
▷ Which areas have been printed, and which left blank?

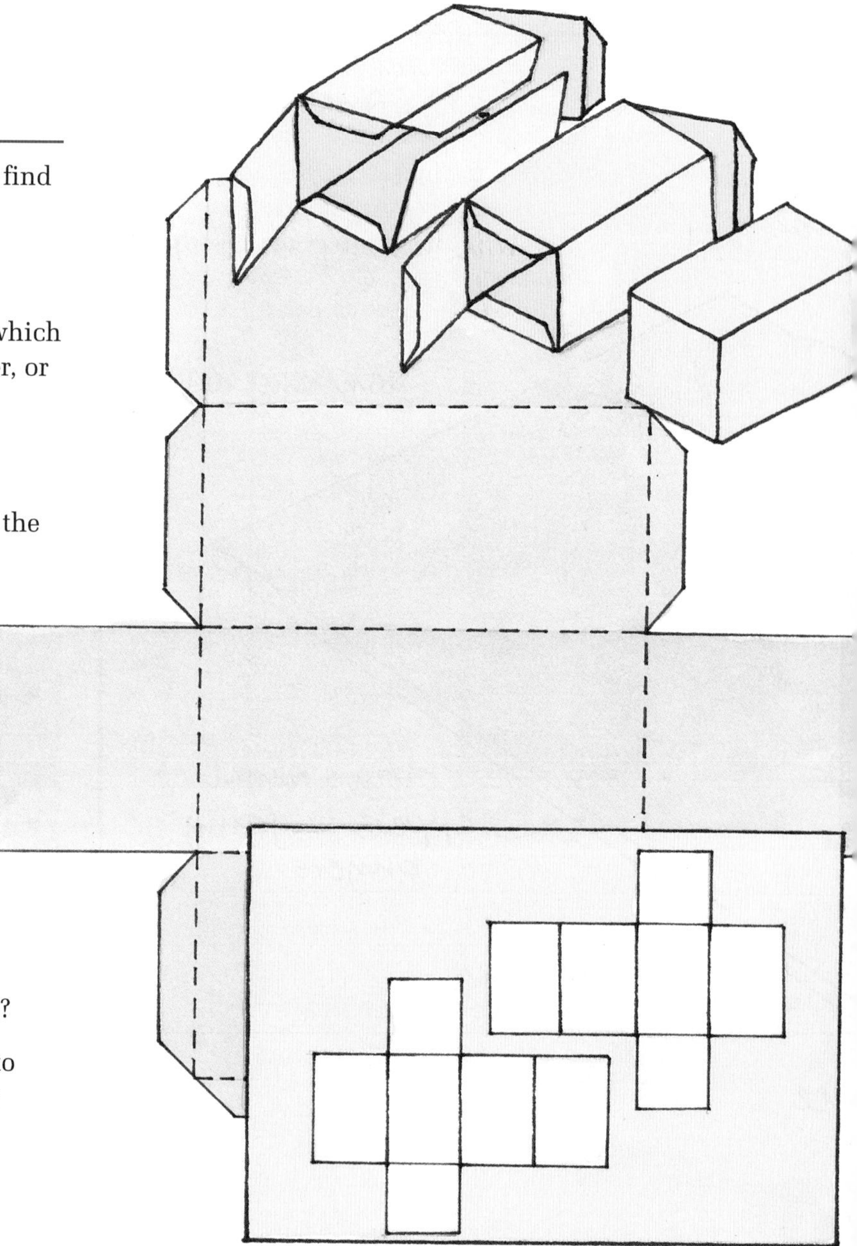

Reducing Waste

When printed, assume the net you have studied would be repeated to fit a sheet of card 2 metres by 3 metres. Experiment to discover the most efficient way of placing the net on the sheet.

▷ What is the maximum number which could be printed at a time?
▷ Roughly how much of the card would be wasted?

Are there any minor changes which could be made to the design of the net (e.g. making the sides longer or shorter) which would mean either that:

▷ more could be printed on each sheet, or
▷ more of the material could be used to make the construction of the package stronger.

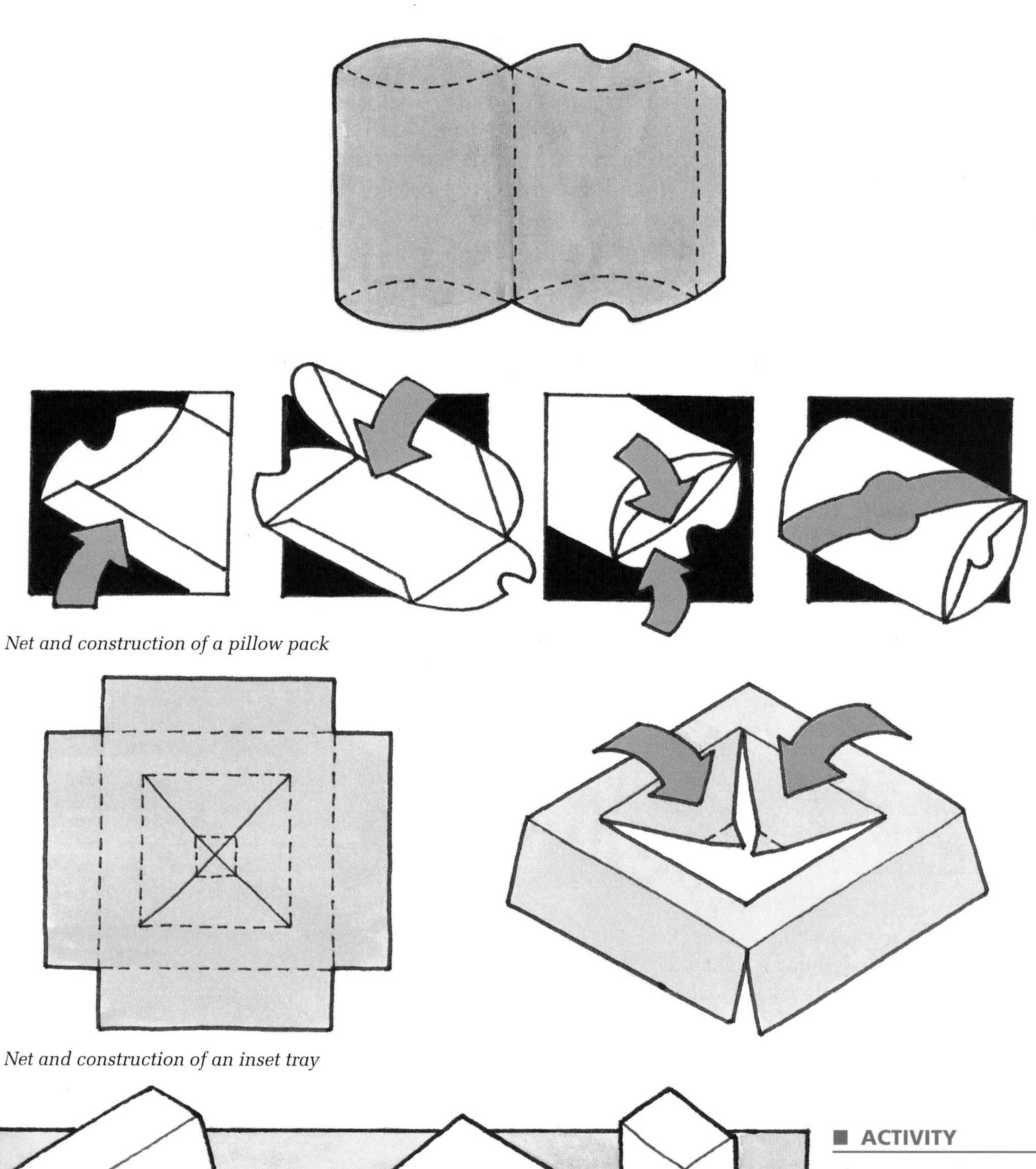

Net and construction of a pillow pack

Net and construction of an inset tray

■ ACTIVITY

Choose a product and select one of the shapes shown opposite. Work out what its net would be like and add appropriate graphics to it.
Finally fold it up into its three- dimensional form.

Designing and Making Moulds

Many plastic packages are made using large industrial vacuum formers. On a smaller scale, vacuum forming enables designers to produce quick and cheap experimental prototypes and appearance models of their ideas which have a sufficient level of realism and detail.

Making a suitable mould needs careful understanding of the processes involved. It must be of the required shape for the design and also be made in a certain way to ensure successful results. The mould must be very smooth with slightly angled sides (an angle of five degrees is usually sufficient). All corners and edges should be rounded or 'radiused'. All these features will help a pleasing shape to be formed and aid the removal of the mould from the plastic sheet.

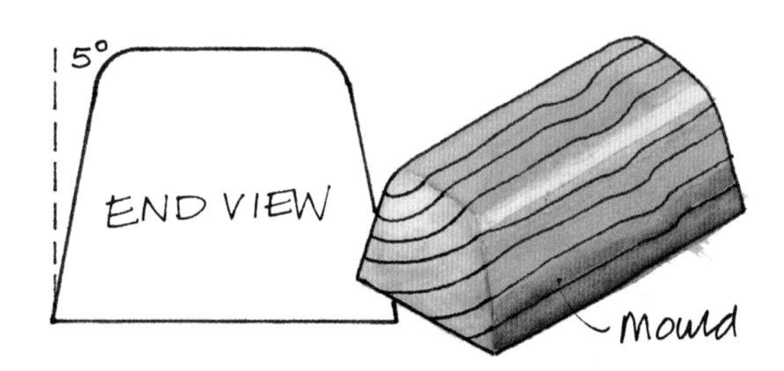

1 The process involves heating a sheet of special plastic in a vacuum forming machine. The supple sheet is then drawn over a pre-made mould (sometimes called a 'pattern' or a 'former') of the required shape. It may be necessary to make the design in two halves and join them later.

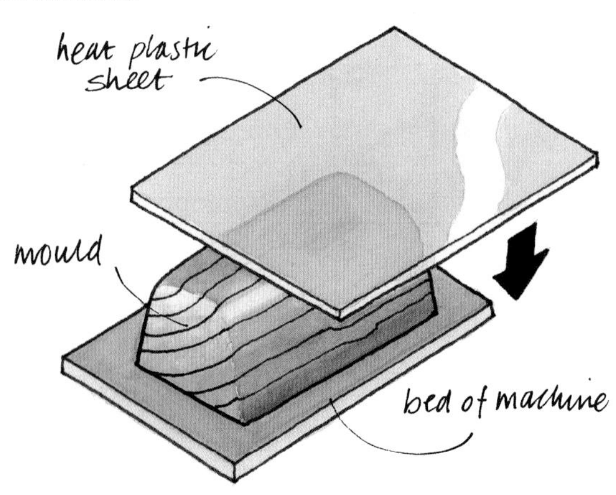

2 A seal is made around the mould on the machine and air sucked out, creating a vacuum. The increased air pressure above the mould causes the sheet of plastic to be drawn down and the shape is formed. The sheet is then allowed to cool, and the shape is retained. This type of plastic has what is known as a 'plastics memory' enabling it to be reheated and returned to its original flat form, if necessary.

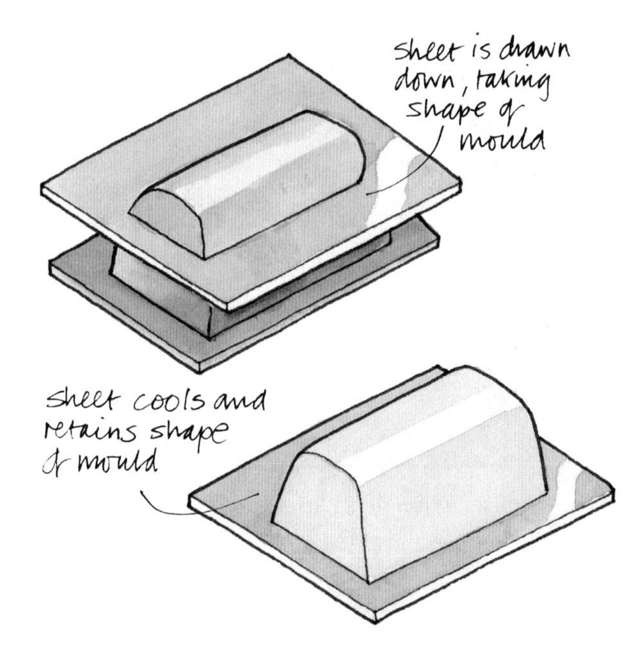

The type of material used for vacuum forming is part of a group known as **thermoplastics**. These can be heated and reformed into a different shape. Vacuum forming plastic (a common one is polyvinyl chloride or PVC) is available in a variety of sizes and thicknesses, and some very bright colours, including transparent, opaque and textured sheets. The plastic does tend to be expensive and care should be taken not to waste it.

3 An amount of curvature of the plastic at the base of the shape is a feature of the process and can be trimmed away. To avoid this, a thin piece of wood such as hardboard can be positioned under the mould, just back from the edges, causing the plastic to 'undercut'.

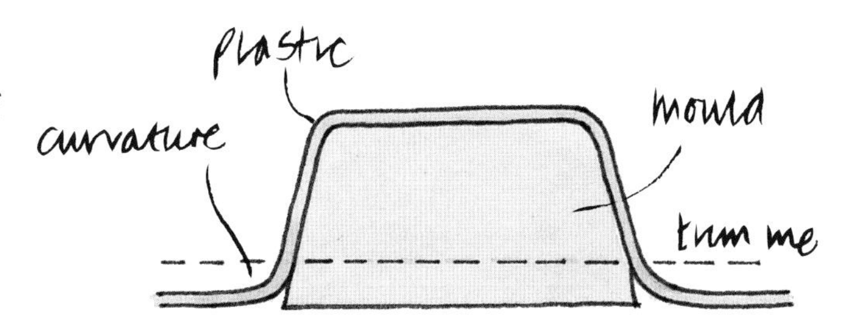

4 The plastic will now need to be trimmed before removal of the mould. Most sheets used can be trimmed with a sharp craft knife, but care should be taken. Unintentional undercutting is often a problem. It occurs when a rushed job has been made of the mould, or existing articles which have complex shapes or recesses have been used as the mould. In drastic cases of undercutting the mould will be impossible to remove without damage to the plastic sheet. In industry much care and attention is paid to the making of the moulds as it is these that determine the quality of the final product and how fast multiples of the design can be made.

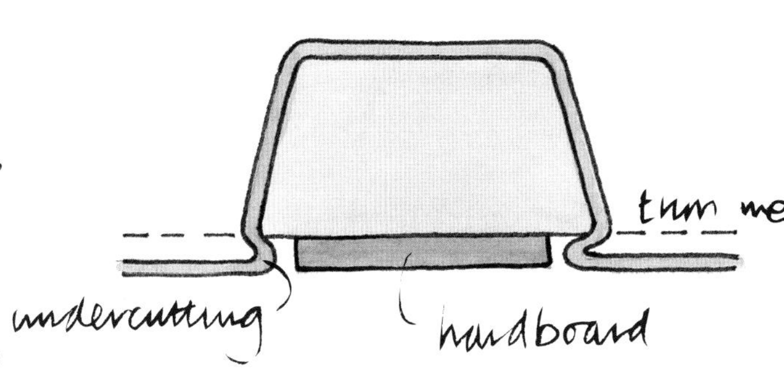

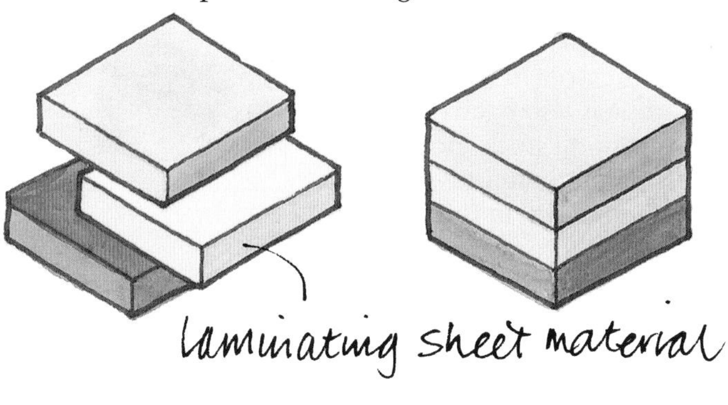

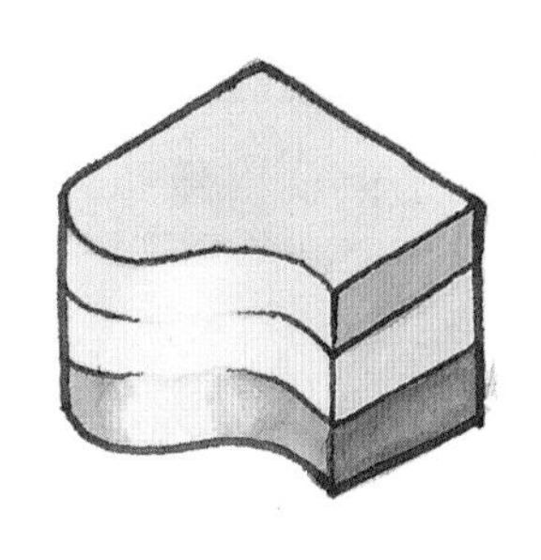

IN YOUR PROJECT

- Work out the details of the shape of the net or mould you need. Experiment in rough first.
- Make sure you check the dimensions to ensure it will be accurate when made up.

A range of different materials can be used to make the mould but woods such as MDF (medium density fibreboard) are ideal. If sheet material is being used, layers can be glued together to give the required thickness, and any blemishes filled and sanded as required.

ACTIVITY

Look carefully at the shapes shown opposite. Choose one and make a suitable mould to reproduce the shape using the vacuum forming process.

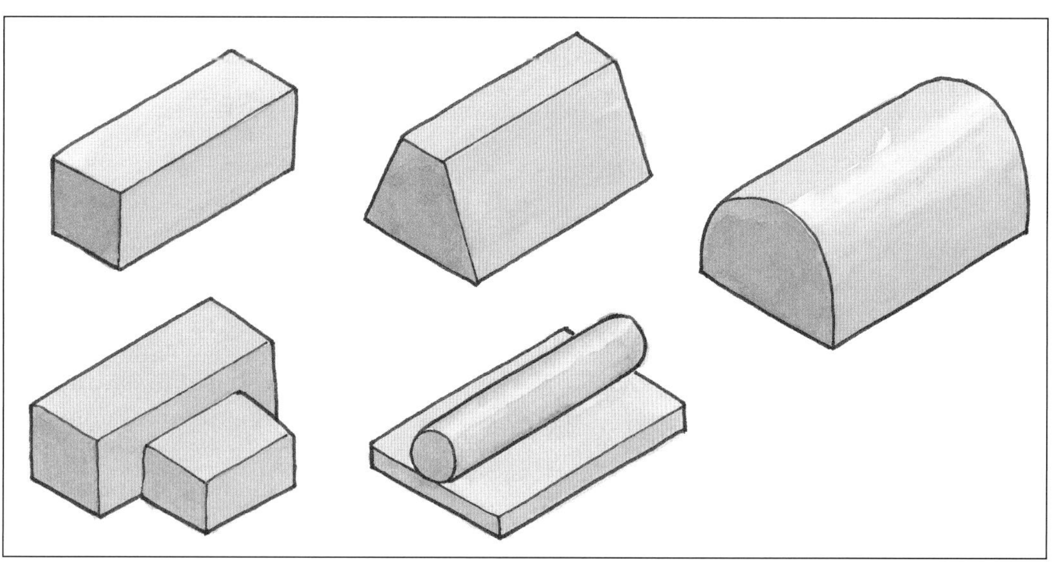

Design Specification

A design brief is a general statement of what the company or a client requires. The designer undertakes research to clarify and add detail to the brief. From this information a list of the possibilities and restrictions of the final design – the design specification – can be prepared.

Writing Your Specification

It is important that the specification is as precise as possible to avoid costly mistakes. It is usually checked and approved by the client before development work proceeds.

There are some important areas to consider in the specification for your Mix and Match package.

- ▷ The container must be large enough to hold the items, but not so big that they rattle around and possibly get damaged.
- ▷ The strength and durability of the packaging materials must be carefully considered to protect the contents, without increasing the cost more than necessary.
- ▷ The container must also be capable of being quickly and easily assembled and filled.

In addition to the selling features on the front of the package there is also mandatory information which has to be included. By law we have to put on its contents, if they are flammable, and the amount of material in the can. A bar -code is also needed for recording stock codes in supermarkets and warehouses. In this instance all that information is placed around the back of the can.

John Walworth, Packaging Designer

Designing by Numbers

Make sure you include some numerical data in your specification. Try to provide data for anything which can be measured, such as size, weight, quantity, time and temperature (see pages 86 and 87).

Package Specification

- ✓ My package will contain...
- ✓ The minimum and maximum height, width and depth should be between...
- ✓ The materials will need to...
- ✓ The items need to be removed without damaging the package.
- ✓ The information which will need to be included on the outside will be...

Designing for Need

Don't forget that your designs will need to be suitable for a range of users who will have a variety of physical needs and psychological preferences. For example, some people have larger hands than others. Pastel colours might be preferred by some users, while others go for bright primaries.

Checklist

Make sure you write some statements about the following:

- ▷ the exact sizes of the items to be packaged
- ▷ how much protection the products will need
- ▷ the weight of the items
- ▷ if the container is to be used to store the items after it has been first opened
- ▷ what product information needs to appear on the package
- ▷ the name of the product
- ▷ what quantities the product will be sold in.

Environmental Concerns

Designers are responsible for ensuring that the products they create will minimise the potential damage to the environment. One way of doing this is to remember to apply the 3 R's:

Reduce

Aim to reduce the amount of:

- ▷ materials being used by making them thinner, lighter or smaller
- ▷ packaging being used to protect products during transportation and storage
- ▷ energy used in the manufacture of products

Re-cycle

Consider whether:

- ▷ the product could be made from recycled materials, e.g. glass, paper, aluminium foil, clothes, etc.
- ▷ it will be easy to recycle the product itself when it is finished with.

Re-use

Try to specify parts and components (e.g. electronic circuit boards, fastenings, packaging, etc.) which could be easily re-used when the product is finished with.

Also think carefully about the maintenance of the product. Will it be easy to repair or replace individual components and parts when they fail, to avoid having to throw the whole product away?

IN YOUR PROJECT

You should try to take into account how your product can be made in the most environmentally friendly way.

You might :

- ▶ decide to specify maximum amounts of some materials
- ▶ avoid a particular material because it can't be easily recycled.
- ▶ specify a manufacturing process because it consumes less energy.

Some of these requirements may conflict with other parts of a specification, such as ease and cost of production.

Good quality pencils are usually painted to aid colour identification and packed in metal boxes. However, Berol's 'Karisma' brand of pencils remain unpainted. Instead they are champfered to aid identification, and to provide a distincive design feature.

Meanwhile the packaging of pencils reflects the natural look of the product. To achieve this the card boxes are wrapped in textured recycled paper to give a hand-craft feel. Embossed, rather than printed, hallmarks express quality.

Threaded eyelets hold the lid in place, removing the need to complex plastic or metal components.

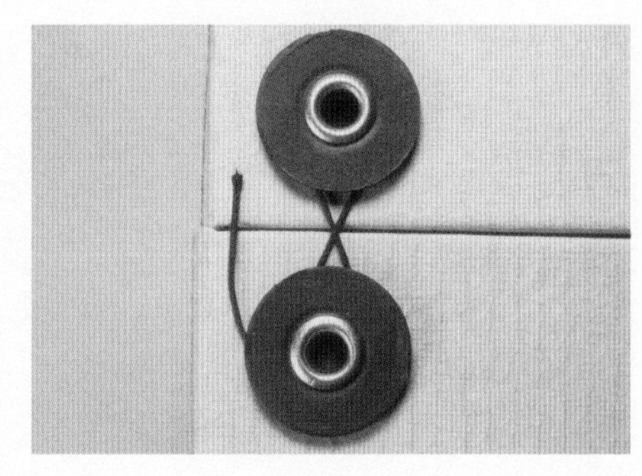

Packaging Materials

There is a wide range of different materials for the designer to choose from for packaging. Each material has its own characteristics. Some are rigid, while others are flexible. Some are made from new materials, while others have been recycled. All of them come in ranges of size, weight, thickness, colour and finish.

A smart material is one which changes its properties and characteristics in a predictable way when under certain conditions. For example, a low melting point thermoplastic which can be shaped by hand after immersion in hot water.

Boards

Paper and card are used extensively in the packaging industry. They are cheap to make, provided the production run is more than 1000 units. Their main disadvantage is that they become weak if they get wet.

Material	Applications	Characteristic
CARDBOARD (300 microns)	▸ High-volume colour printed cartons	▸ Low strength/ weight ratio ▸ Excellent printing surface ▸ Recyclable
CORRUGATED CARD	▸ Protective packaging for fragile goods ▸ Low-cost protection	▸ Recyclable

paper
card
corrugated card
aluminium foil
fibre board
plastic bubble wrap
chipboard
foil lined board

Board Types

There are many different types of board to choose from. Stronger or specially treated boards are more expensive to use.

Material	Applications	Advantages
DUPLEX BOARDS: (pure wood pulp) unbleached body plus a bleached liner (350-640 microns)	▸ Tobacco, food and pharmaceutic packages ▸ Provides textured printing surface	▸ Good for printing high-speed automatically-packed cartons
SOLID WHITE BOARDS: pure bleached wood pulp (350-640 microns)	▸ Book covers, cosmetics cartons	▸ Very strong ▸ Excellent printing surface
MEDIUM DENSITY FIBREBOARD: high content of recycled paper and board (2-50mm)		▸ Lower cost ▸ Higher strength/ weight ratio, ▸ can be lined ▸ good for long-run printed graphics
CAST-COATED BOARDS: a heavier and smoother coating applied to duplex and solid white board	▸ Luxury products with expensive-looking decorative effects	▸ Gives higher gloss after varnishing
FOIL-LINED BOARDS: foil can be laminated to all the above board types (350 microns)	▸ Cosmetics cartons, pre-packed food packages	▸ Can be matt or gloss, gold or silver ▸ Strong visual impact ▸ Provides a barrier

Glass

Glass packaging is used mainly for liquids such as drinks and perfume, and for jams and other food products with a liquid content.

Material	Applications	Advantages	Disadvantages
GLASS	► Liquids, e.g. perfumes, drinks, jams.	► Low cost ► Contents are visible ► can be coloured and moulded in unusual and distinctive shapes	► Heavy ► Breaks easily and shatters on impact ► Expensive tooling required ► Disposal problems
METALS	► Food and drink cans ► Aerosol cans for deodorants, spray paints and household cleaners	► Metal may be printed on when flat or after forming	► Printing onto metal is more expensive
PLASTICS	► Liquids such as detergents and lotions	► Cheaper and lighter than glass ► Easier to form into complex and unusual shapes and textured surfaces than glass	► Less environmentally-friendly than glass ► Disposal problems

Metals

Aluminium and tin plate are used to make cans. Graphics are usually printed onto a label which is wrapped round and stuck on the can. For special effects it is also possible to print directly onto the tin.

Plastics

Plastics can be made clear or opaque in a variety of colours. It is cheaper and lighter than glass, and easier to form into complex and unusual shapes and textured surfaces. Heat and pressure are applied to the plastic to produce the shapes required. There are many different types of plastic. They are usually called by their initials:

- ▷ LDPE: low-density polyethylene
- ▷ HDPE: high-density polyethylene
- ▷ PP: polypropylene
- ▷ PS: polystyrene
- ▷ PET: polyethylene terephthalate (polyester)
- ▷ PVC: polyvinyl chloride.

IN YOUR PROJECT

When choosing suitable board, you need to be sure that:

- ► it is suitable for the intended printing process
- ► it will bend and fold without damage
- ► it is suitable for gluing on automated gluing machines
- ► it will not collapse when used on automated packaging equipment.

KEY POINTS

- Card, glass, metal and plastics are all used for packaging.
- Each has its own advantages and disadvantages in terms of recyclability, cost, strength and suitability for printing.

Green Packaging

The packaging industry uses a high volume of natural resources, such as trees, oil, sand and metal ores. Manufacturing and processing of packaging materials also requires energy. At the end of the day, however, most packages are simply thrown away.

Packaging designers need to think very carefully about what will happen to their containers after they have been finished with.

What a load of rubbish!

A third of all rubbish is packaging materials! Most of the things we throw away end up in land-fill sites, buried into the ground. There they last for anything between two weeks to one million years!

These environmental concerns are being addressed by the packaging industry in response to Government legislation. Many packages are now being made as thin as possible to conserve resources and energy. Others are being made from **biodegradable** products, such as untreated paper, which decompose much more quickly. The aim is to create packaging which still promotes and protects the product but which is less harmful to the environment.

Recycling waste packaging is another approach, although sometimes this is expensive in terms of energy and in the cost of collection and sorting.

The current demand for packaging can only increase as our lifestyles demand the convenience it brings. This may become a serious problem for future generations. Packaging designers must strive to minimise pollution and the use of resources. Meanwhile, thinking carefully about how we dispose of our waste packaging materials is also essential.

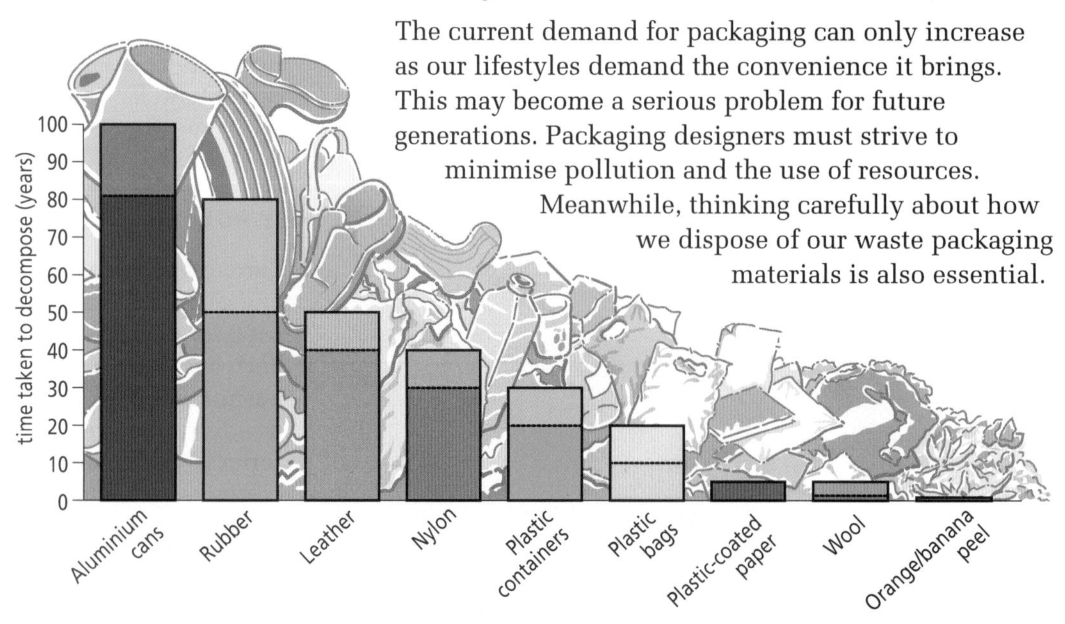

■ ACTIVITY

Read the various approaches to the problems of producing more environmentally-friendly packaging, shown on the right.

Find two examples of packaging which use one or more of these approaches. Draw them in colour on an A3 sheet, adding annotation to explain their design.

Green Approaches

Less is More

One approach is to consider whether the amount of packaging being used for a product is really needed. Confectionery and cosmetics blister-packs are some of the most wasteful. The most creative design work often uses simple materials such as corrugated card and printed paper to achieve subtle and sophisticated packages. They are also simpler and cheaper to produce.

Is all this packaging really needed?

Reducing the amount of material used in surface area and thickness is an obvious approach, particularly where the package will only be used once of for a very short time.

The shape of the pack can make a difference to transport costs – square containers are much more efficient as they contain more product and less space when packed together. Striking graphic images and typography, unusual colour schemes and textured papers can compensate for a conventionally shaped box.

Refill

Reusing and refilling containers is an obvious way of reducing packaging costs. It is only suitable for some products, however. The expense and energy consumption involved in returning the container to the manufacturer and cleaning them fit for re-use can be costly.

Bottles are well suited to this approach as they can be used many times over. Another method is for the consumer to take their own container and fill it directly from a larger storage unit.

With a refill pack, one durable container is purchased initially and then thinner, lighter packages can be used for subsequent purchases. Refill packs need to be easy to use, so pouring spouts and easy-to-open fastenings are important.

Recycling

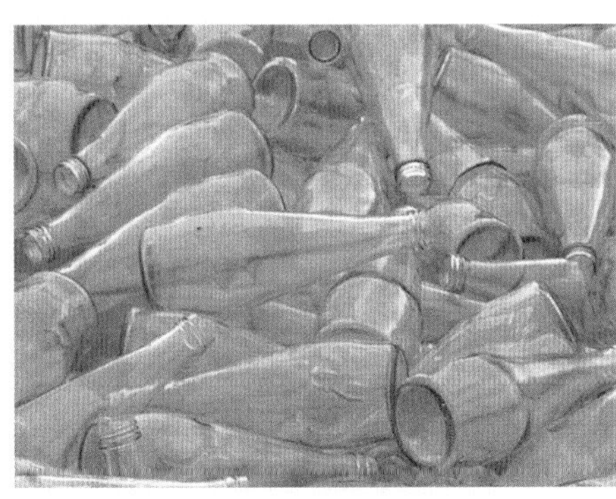

Glass is probably the easiest packaging material to recycle and use again for the same purpose.

Often recycled papers and cards, rather than newly pulped and treated materials, are adequate for the job. Recycled papers and boards do have different working characteristics, however.

Many types of plastic can be recycled but their chemical structure is weakened. As a result they may not be able to take up more complex shapes with precision.

Biodegradability

Biodegradable materials are those which decompose relatively quickly. Unfortunately this means that they also only give limited protection to the product. Biodestructable plastics is a technology still very much under development, but one day it might be possible to produce materials which decompose more rapidly than at present.

Hazardous Materials

Some packaging materials may be potentially damaging to health and the environment during production, in use and when eventually thrown away. PVC and CFC gases are of particular concern and should be avoided where possible.

Litter

Reducing the number of separate parts in a package helps reduce the possibility of them being 'dropped' rather than thrown away. One familiar, but now solved, problem is that of the can ring-pull. How has this been solved?

Packaging designers need to graphically remind and prompt users to throw packaging materials away properly.

Packaging Ideas

You should now be able to develop your design ideas for your Mix and Match container.

Is it going to be a cardboard container or a vacuum-formed bubble pack? What are the most environmentally-friendly materials it can be made from? How will it be printed?

Designing a Container

Experiment with different nets to discover the one which best protects the products you have chosen. (See page 60.)

Designing a Bubble Pack

Make up a mould which is slightly larger than the products you are going to contain. (See page 62.)

Developing the Graphics

Explore ideas for colour and typography, based on a grid design. (See page 36.)

Specifying the Printing

Work out how your package might be printed. What finishes or special effects could be used? (See pages 71 to 73.)

Organising the Production

In what order will the package be made? When will the quality be checked? (See pages 74 to 77.)

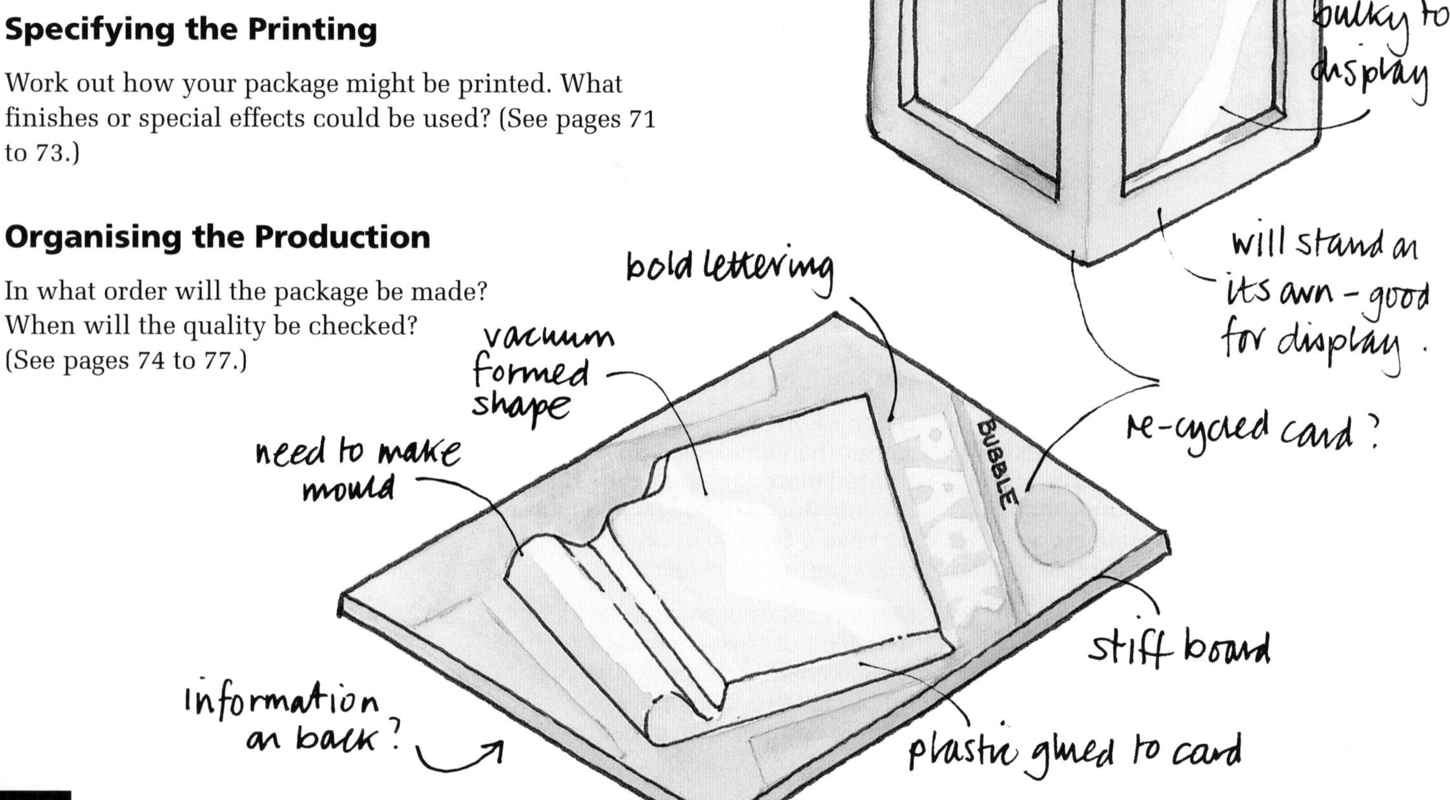

Paper and Ink

There are many different sizes, colours and weights of paper, card and board for the graphic designer to choose from. There is also a wide range of types of ink. It is important to specify the correct ink to produce the required image on the chosen surface.

Paper

Paper is the most versatile of all printing materials. Paper is made by machine from pulped wood, and is available in a huge range of thicknesses (or weights), colours, types, textures, and sizes.

Papers are sold in weights, usually grams per square metre or gsm. An expensive publication, such as one of the better magazines, might use a paper of 150 gsm for its cover and 85 gsm for the inside pages. The surface of the paper may well be of a glossy appearance and bright white in colour. In contrast, most newspapers use thinner, matt paper which is very absorbent, off-white in colour and very much cheaper.

Coloured papers are particularly useful for backgrounds but can be used as a medium for design in their own right. An extensive range of coloured papers and surface finishes and textures are also available. It is possible to buy coloured paper that is exactly matched to the colours that will be specified to the printer. Pantone supply a range of cards for this purpose. The designer is able to produce visuals for the client with the confidence that the finished printed item will be the same.

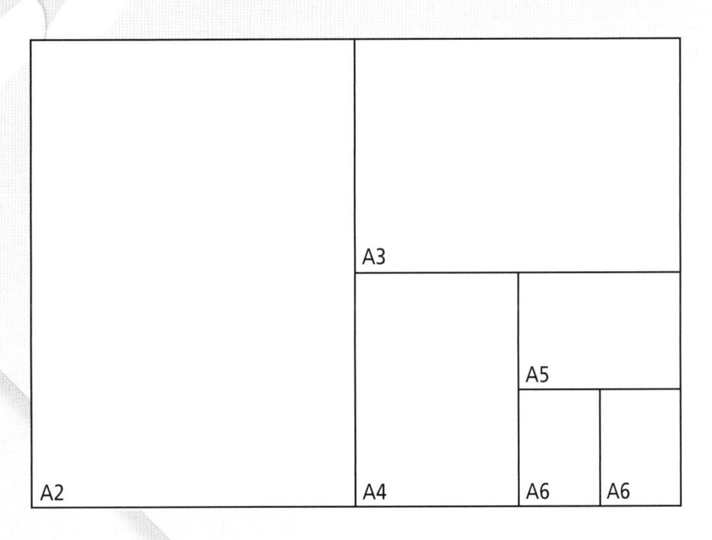

The common paper sizes are known as 'A' sizes. A4 is the most common size of paper found in schools and offices. A3 and A2 are the most useful sizes for designing on. Most paper used in design and technology departments in schools has a matt finish and is about 80 gsm in weight.

■ ACTIVITY

Contact a paper supplier or printer in your area and try to obtain a book of paper samples.

Make a collection of samples of different types of household paper, such as stickers, labels, packages, brochures, etc. Stick a selection of them onto an A3 sheet and describe their weight, texture and ink.

Inks

Modern inks are made using complicated chemical processes. To work they must remain liquid to aid transfer to the paper and then dry as quickly as possible. Inks are made up of a solid pigment (the colour) and a liquid part (called the 'vehicle').The vehicle is either absorbed into the material being printed or evaporates.

Into Print [1]

Graphic products can be reproduced in a variety of ways. The printing processes selected by the designer will depend on the length of the print-run, the nature of the artwork, the quality of paper and print needed, and the overall cost.

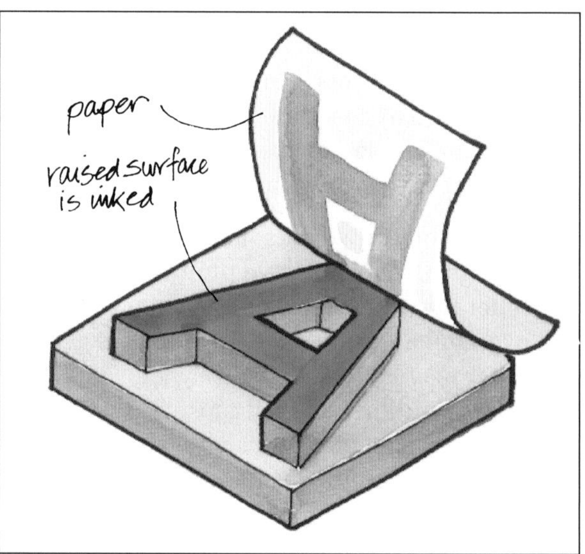

Letterpress printing

Printing Methods

There are four main methods of printing. Each has its own advantages and disadvantages.

Method	Applications	Advantages	Disadvantages	Print Run	Cost
LETTERPRESS	► Books with large amounts of text, letterheads and business cards	► Good quality in terms of sharpness and solidity	► Few suppliers ► Not economical for colour work ► Slow process	► 500 to 5000 copies	► Only competitive for short runs and reprints of monochrome work
LITHOGRAPHY	► Newspapers, magazines, books, posters, letterheads and packaging	► A widely used modern process ► Many suppliers ► Good print quality	► Plate life limited to 150 000 copies, but can be quickly remade	► Ideal for medium to long runs: 250 000 to 1 million copies	► Most economical process for general printing
GRAVURE	► Expensive magazines, books and postage stamps	► Produces very high-quality print work	► Expensive to set up	► 500 000 to several million	► Very high print runs required to recover set-up costs
SCREENPRINTING	► Posters, T-shirts, shop display boards, fabrics and wallpapers	► Can be printed onto smooth, absorbant or rough surfaces	► Only suitable for short runs ► No fine detail possible	► Good for short print runs: several copies to a few hundred	► Low cost

Letterpress

Letterpress is one of the oldest forms of printing, perhaps invented by the Chinese who used wooden blocks to print from. The first books to be printed used this method and were produced in the 14th century. Johannes Guttenberg, a German, invented moveable type and the first printing machine. The first English printer is thought to have been William Caxton.

Moveable type meant that the individual pieces of type could be arranged into words and sentences. Wooden blocks were carved to provide illustrations for the text. In more recent times machines were invented to cast the text in lines and blocks saving the need to compose the type by hand.

Letterpress printing is a form of **relief printing**. This means that the image area (the area to be printed) stands above the non-image area (the area not to be printed). The image area is inked and the image transferred to paper.

Letterpress printing was still widely used until about 15 years ago. Today its use is mainly restricted to small printers who economically print items such as letterheads and business cards. It is an ideal process for small print runs and single or two-colour work using standard typefaces.

It is especially suitable for producing work with large amounts of plain text, such as books. It has limitations for the graphic designer as the range of typefaces and flexibility of layout are extremely limited.

The quality of letterpress is, however, extremely high in terms of sharpness and solidity. It is also possible to make last-minute detail changes to a page without having to make a complete new plate.

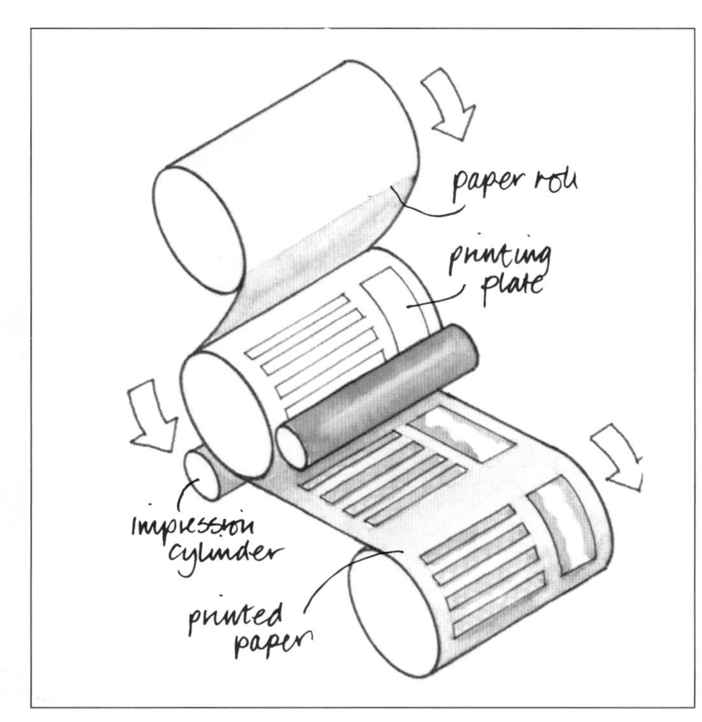

Rotary letterpress

Rotary letterpress is a modern relief printing process using a thin plate usually made of an aluminium base sheet with a thin covering of plastic. The plastic has the image transferred photographically onto it and the unwanted non-image areas are removed by a solvent leaving the final relief plate.

Rotary letterpress is very economical for text and line work which will be frequently reprinted. The process is usually used for bookwork and some packaging. It is not suitable for half-tone pictures, colour or pictures with fine detail.

Lithography

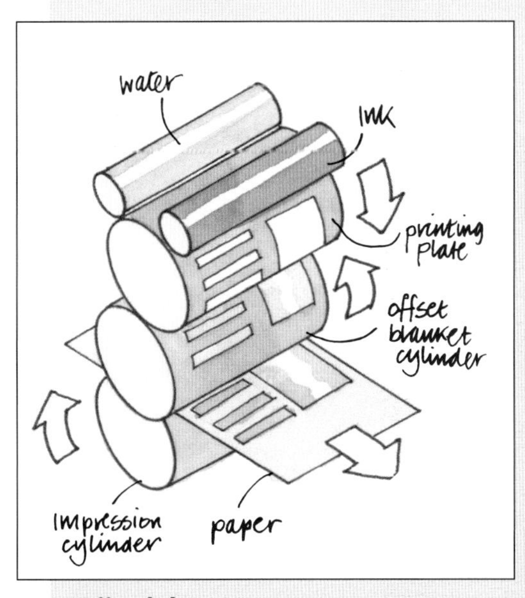

Offset litho

Lithography dominated the printing industry in the latter part of the 20th century, although it was first introduced in the 1700s.

The process relies on the principle that oil and greasy substances do not mix with water. The printing plate, originally limestone, is now made of thin aluminium sheet and is totally flat unlike a relief plate.

The image is transferred to the plate photographically by the use of ultra violet light shining through a negative. The plate is washed with a chemical that makes the image area attractive to oil or greasy substances, i.e. the ink. The non-image area is later dampened with water which will repel the ink.

Lithography has widespread use and is ideal for medium to long print runs because the setting-up costs can be expensive for short runs. It is used for single or multiple- colour work such as magazines, posters, letterheads, packaging (including CD booklets) and tickets.

Offset litho is a term referring to the method of transferring the inked image from the plate to a rubber blanket and then onto the paper. This avoids direct contact between paper and plate, which lengthens the working life of the plate and prevents the paper from becoming damp through contact with the water used during the printing.

Web offset litho refers to the paper being fed from a web or reel instead of being sheet fed. Newspapers and books are printed in this way.

The longer the run, the lower the unit cost. For a single-colour letterhead, 100 copies might be economical, but for a four-colour process run, up to 250 000 copies must be printed for the process to be economical.

Into Print [2]

Gravure

This is a high-speed rotary process which is used for high-quality print work. It is expensive to set up and is only viable for long print runs over 500 000 copies.

Gravure is typically used for illustrated magazines, colour newspaper supplements, mail order catalogues and long-run colour commercial work. The process uses a form of etched plate where the printing surface is lower than the main area of the plate. Ink is applied to fill the image area cells and the surface is scraped off. The paper is brought into contact with the plate surface by means of an impression cylinder, and the image is drawn out of the plate surface.

Heavy copper-faced printing cylinders are photo-etched to create the image and attached to the press. This is expensive, but provides a durable image over a long print run.

Sheet-fed gravure is used for specialist printing of fine art prints, stamps and some packaging.

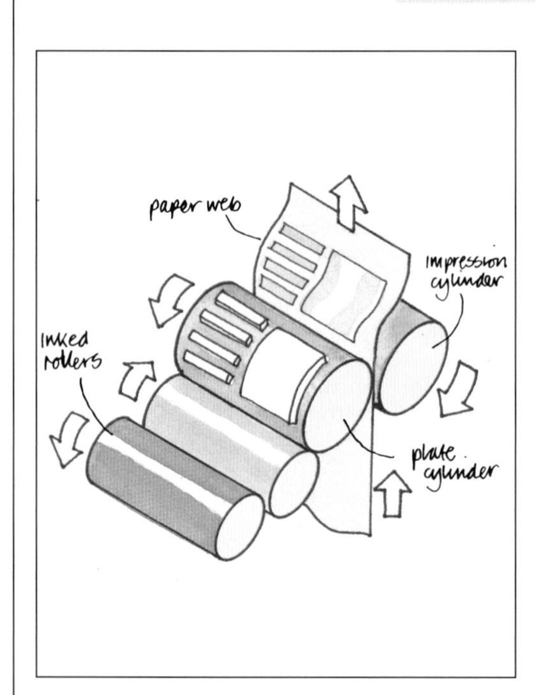

Flexography

As a process, flexography is similar to letterpress in that it uses a relief plate. As the name implies, the printing plate is flexible, usually being made of plastic or rubber.

The process is used for printing many types of packaging, often when the material to be printed is not conventional paper (say to print polythene used for food wrappers). It is particularly good on these surfaces because they are very thin and non-porous. The type of ink used for flexography is often mixed with a solvent which evaporates on contact with the polythene and enables the ink to dry quickly. It is suitable for very large print runs and is a very speedy process. Items printed include carrier bags and some wallpapers. It is particularly suitable for frozen-foods packaging and multicolour corrugated point-of sale displays.

Activity

See if you can find out more about the special techniques needed to print:

- self-adhesive vinyls
- cardboard tubes
- tin cans
- bank notes
- bank cheques
- credit cards
- CDs
- glass bottles
- gift wrapping paper
- T-shirts
- billboards.

Screen printing

Screen printing is completely different to other types of printing process. It relies on a stencil through which the ink is forced. The holes or gaps in the stencil correspond to the image area to be printed.

The stencil can be made of a variety of materials that include paper, thin card, and special photographic materials. They are supported on a fine mesh and it is on this that the ink is spread and scraped along forcing it through the mesh and the stencil and onto the surface of the printed article. Simple designs work best with paper or card stencils, however much finer detail can be achieved with a photographic stencil (these are usually made up professionally).

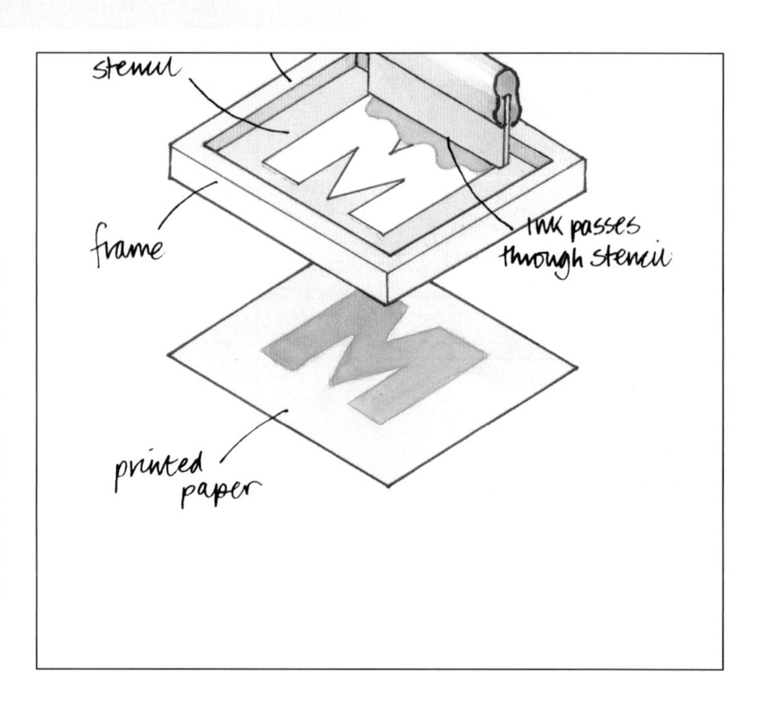

Colour Printing

A great deal of commercial work is in full colour. The designer needs to be aware of the process of full-colour printing.

Modern colour printing uses a dot structure made up of the three primary colours and black to create the full range of colours when superimposed on the page. These are known as process colours (see also page 34).

A separate printing unit is required for each colour. Modern four, five or six colour fast litho printing presses print each colour in sequence with additional units adding special colours or finishing coating (see page 76) as an in-line process.

A camera or a scanner is used to separate the red, the blue and the yellow from the image. In printing terms these colours are known as magenta, cyan and yellow. Black is also separated and four individual pieces of film made from which four printing plates will eventually be produced. The separation is achieved by exposing the film through coloured filters.

A screen is also used that converts the separated colours into dots. The dots are very small but can be clearly seen if you look closely at a photograph that has been reproduced in a magazine. The process of positioning one printing exactly over another in this way is known as **registration**.

When using process colours it is difficult to obtain an exact specification for a particular colour. This may be essential for corporate symbols and logos or product packaging. In particular golds, deep blues and some greens can be difficult to match.

It is more reliable to print pictures from the four process colours and use additional units for special colours.A range of standard colours used for identification in the trade by graphic designers and printers is known as Pantone. The Pantone colour range comprises 700 standard shades, each defined by their percentage of the four colours (e.g. 30% cyan, 0% magenta, 65% yellow, 5% black).

Accurate colour-matched proofs guide the printer to control the process and can be checked and approved by customers before a long print run.

Process colour printing is much more expensive than printing in one colour only.A cheaper option is to restrict the number of colours and use them creatively for emphasis, contrast, or special effect. These are known as spot colours.

IN YOUR PROJECT

- Make a statement about which of the four main printing methods would be best for printing the graphic product you have designed.
- If it is in colour, provide a brief explanation of the printing process, and precise details of any specific colours to be used.

KEY POINTS

- The four main printing processes are letterpress, lithography, gravure and screen printing.
- The four process colours are magenta, cyan, yellow and black.

Printing Effects

There is a range of special printing techniques a designer can use to help add interest and impact to a publication. These usually involve using different combinations of papers, inks and varnishes. Some of them add considerably to the cost of printing however.

Special Effects

Effect	Applications	Advantages	Disadvantages	Cost
DIE CUTTING	► Packaging ► Unusual-shaped papers	► Once set up, cutting die can be reused many times	► Slow process	► Expensive for short runs if special shapes are required
SPIRIT VARNISHING	► Protects paper and card ► Looks more attractive	► Easy process for enhancing or protecting product from scuffing	► Cannot be added until printing ink is dry	► Low cost ► Can be applied on the printing press
UV VARNISHING	► A heat-cured coating which uses ultra violet light ► Enhances magazine and book covers	► Increases durability ► Provides high-gloss finish	► Does not add any strength to the product	► Half the cost of laminating ► Can be applied at the same time as printing
LAMINATING	► Cartons, special brochures and company reports, menus	► Increases durability ► Provides high-gloss finish	► Expensive ► Can peel and blister	► Twice the cost of varnishing
EMBOSSING	► Special effects for packaging, business cards, etc.	► Attractive, enhances graphic designs	► Expensive process requiring special tooling	► Doubles initial print cost

Die Cutting and Folding

Paper and card with straight edges are cut to shape and size in batches on large guillotines. Die cutting is a machine process where a quantity of an irregular-shaped design can be produced, providing the paper is not too thin. This includes cutting the outside shape and any holes within the design.

Using a CAD/CAM system, a special-shaped blade is made for the machine, rather like a pastry cutter, and the design stamped out. This method is used extensively in the packaging industry to cut out the net of a design. Laser cutters enable sample prototypes to be made and evaluated quickly and cheaply.

Any folds in the design are achieved in the same way except that the blade does not cut right through the design but merely squashes it against a shaped recess, forming a line along which the fold is later made.

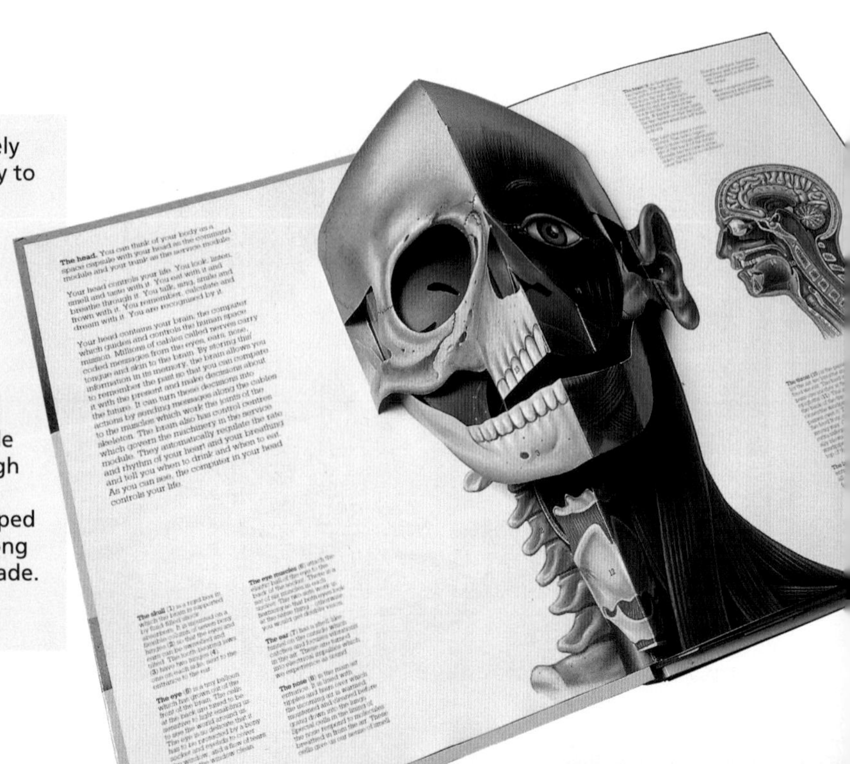

Varnishing

Applying a thin glossy varnish to a printed product helps protect it at the same time as making it look more attractive. Varnishing requires a further stage on the printing process after all the colours have been printed, but before any necessary folding, cutting and trimming.

There are four types of varnish:

- oil-based, which takes between two and eight hours to dry, and is the cheapest to use
- water-based, which involves the use of a special machine
- ultra-violet, which are highly smooth and glossy and dry almost instantly, but require that the ink must be completely dry: this adds cost and time to the production process
- spirit-based, which are now environmentally unacceptable.

Varnishes cannot be added until the ink has dried, unless special ultra violet inks are used.

Spot varnishing is a method of printing a layer of varnish over a specific part of a printed surface to provide a dramatic highlight. The glossy varnish is usually applied over colour photographs or graphic shapes to create a contrast with the surrounding matt background.

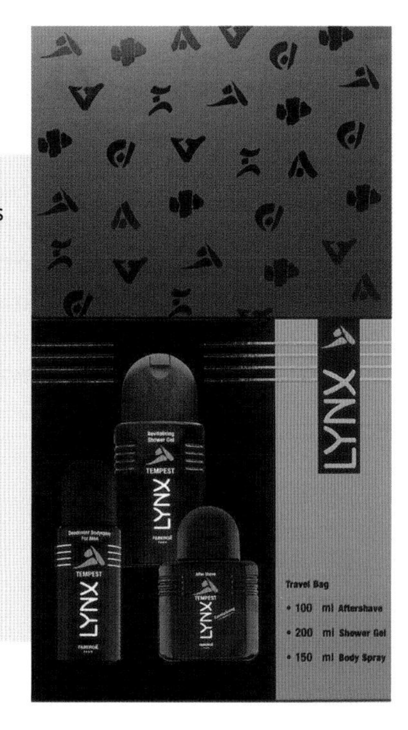

Laminating

Laminating is similar to varnishing. It provides a high-gloss finish which also increases the durability of the printed product. It achieves this by bonding a thin plastic coating over the entire printed surface after printing. Laminating is more than twice as expensive as varnishing, but does provide a glossier finish, and much greater protection. This is important if a large number of people are likely to handle the product.

Lamination is used mostly for products such as cartons, short-print runs of brochures and company reports, some paperback and magazine covers, and menus.

It is possible to laminate one-offs such as presentation artwork, but this tends to be quite expensive. It can, however, be worthwhile for the quality of the finish achieved.

Embossing

Embossing is the process by which a portion of the surface is raised above the surrounding area. It is done for effect, both visual and to the touch, and provides a very subtle look and feel of quality. The effect is enhanced when used in conjunction with carefully chosen typography and textured papers.

The required shape is pressed into the printed work using a steel die. Embossing involves a separate stage in the production process, after any varnishing and laminating. It costs about the same as printing, so if a design includes ink and embossing it will cost twice as much.

Most types of paper and board can be embossed, and there are no restrictions on size. Embossing is a popular for products which need to look expensive, such as high-class letterheads, business cards, luxury product packaging and occasionally for the covers of paper back books.

It is possible to achieve one-off embossed effects for presentations by cutting a shape out of stiff card or thin wood and clamping this together with the image.

Other printing effects include the use of metallic papers (called 'foil blocking') and inks, and holography. See what you can find out about these.

IN YOUR PROJECT

- Consider and specify the use of special printing effects for your product.
- A very important consideration is whether the length of the print run justifies the expense of a special effect.
- Don't forget to include the extra production processes involved in these when discussing the overall printing costs and time-factors.

Systems and Control

Systems are central to the management and operation of many industrial and commercial organisations. It is important to be able to understand how systems work, how well they work and to make suggestions for ways in which they might be improved.

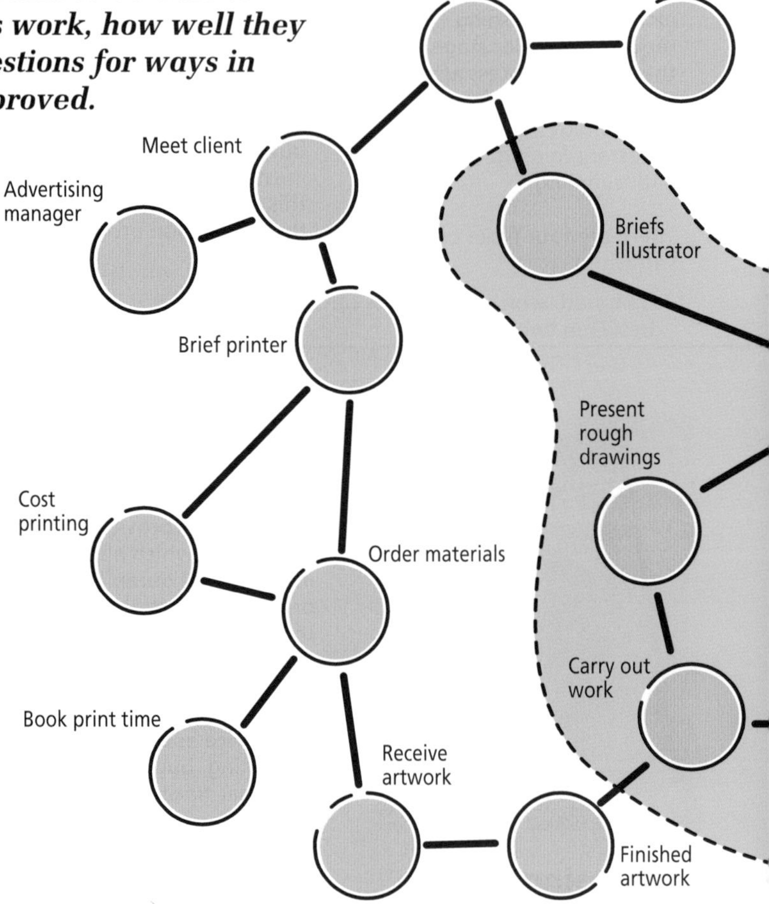

Systems boundary diagram

What is a System?

People often talk about things like 'one-way traffic systems', 'information systems', 'eco-systems' and of the 'solar system'. But what exactly is a 'system'? One important clue lies in the word 'connection'. A system is not so much about the objects and/or events involved, but more about the way in which changing or removing one part of a system alters the other parts.

- ▷ Analysing a system therefore involves identifying its structure and the changes which take place when one or more of the parts are altered.
- ▷ Evaluating a system involves making a judgement as to how effectively it continues to perform while the changes are taking place and what the risks of complete failure are.
- ▷ Designing a system involves creating something which will continue to operate successfully as the elements of the system change. It is also important to ensure that if the system fails, it will not have unacceptable consequences.

Types of System

There are a number of different sorts of system:

- ▷ Natural systems – such as animals, the weather, eco-systems.
- ▷ Designed systems – such as a telephone switchboard, a sewing machine, a bicycle.
- ▷ Abstract systems – such as a computer programme.
- ▷ Human activity systems – such as a school timetable.

Most systems involve combinations of these different types of system.

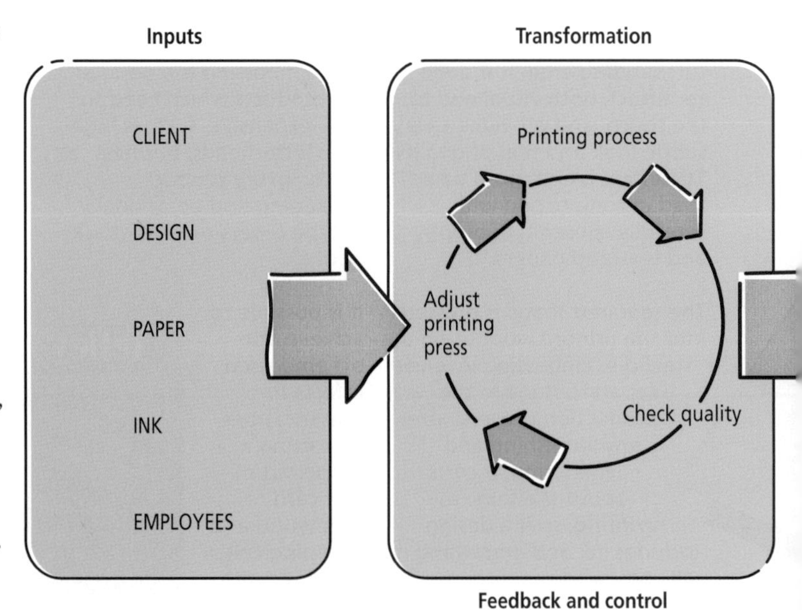

Systems process diagram

System Boundaries

All systems are parts of an infinite number of larger and smaller systems. It is therefore essential to begin by defining the particular boundaries of the system to be studied. Suppose you want to study a printing production system. Are you going to consider the whole printing industry which it is a part of, or just the mixing of different coloured inks?

Inputs, Outputs and Transformation Processes

All systems have 'inputs' and 'outputs'. The main purpose of a system is to change or transform the inputs into outputs. Most systems have many different sorts of input and output, and the first stage in analysing a system is to identify the inputs, outputs and the transformation processes involved.

Again, if we use a printing works as an example of a system, the inputs might include page designs, paper, ink, printing presses and employees. The transformed outputs might include printed products, financially rewarded workers and satisfied clients. The printing works is a system for bringing such transformations about.

Another example might be a take-away restaurant. The inputs might include hungry customers, uncooked food, packaging materials, financial investment, etc. The outputs might include satisfied customers, take-away meals, profit or loss, etc. The restaurant is a system for transforming these inputs into outputs.

It is also possible to analyse a system in terms of a sequence of events, e.g. order food – pay – carry to table – eat – clear table.

Some transformation processes serve to maintain the 'equilibrium', or balance, of the system. Others work to improve the quantity and/or quality of the outputs. It is therefore possible to identify and analyse whether the various processes going on are maintaining the balance or attempting to improve quantity or quality.

Feedback and Control

When undertaking a systems analysis it might be discovered that the quantity or quality of the outputs are unsatisfactorily in some way – poor quality print for example, or a lack of profit in a restaurant perhaps.

As a result it may be found desirable or necessary to change the inputs, or to alter the process of transformation. This is known as **feedback**. The means by which the inputs or processes are changed are called **controls**. The success of a system is judged by considering how well it transforms its inputs into outputs, and how well it is prevented from failing to work satisfactorily as a result of its feedback and control mechanisms.

In the printing works you might be examining the efficiency of a machine at printing a particular colour over a large surface area. It might be discovered that the acceptable limits of the consistency of the colour are plus or minus 1% over each 1000 sheets. Below this figure the difference in colour is noticeable, and above the limit increases the cost too much. The question then becomes how to ensure the machine works within these limits?

As well as quality of product or provision of a service, systems analysis often focuses on achieving acceptable time delays.

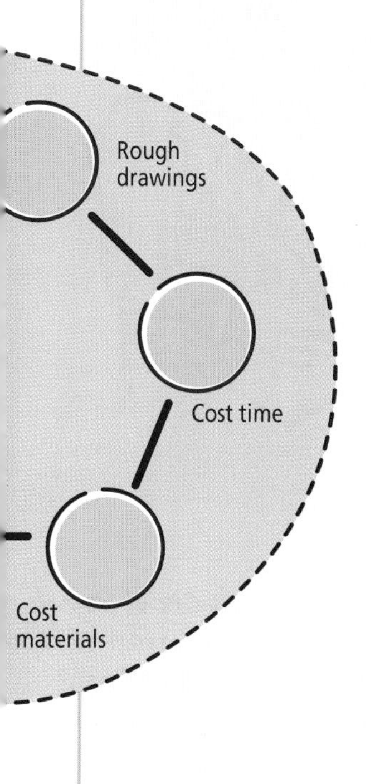

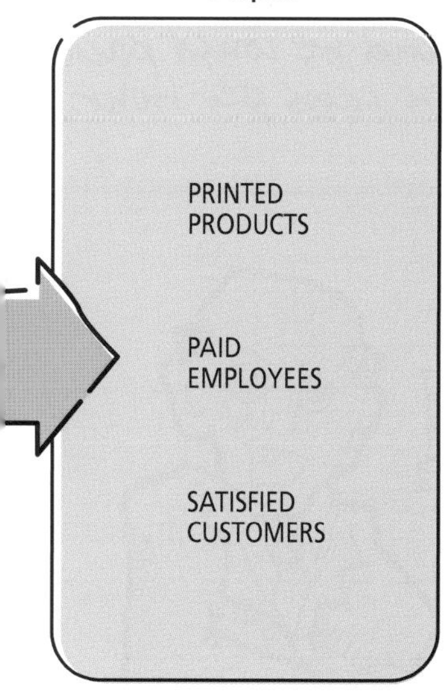

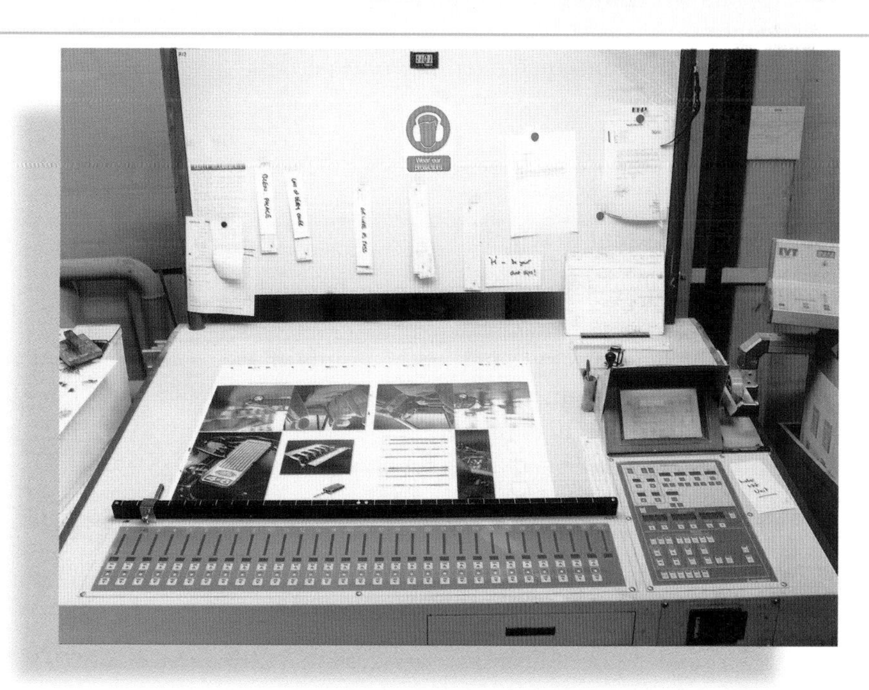

Systems Analysis

Analysing a system involves looking at a complicated situation and being able to identify some degree of structure and connection between the things which are going on.

Stage One

Identify and state clearly:

▷ the main inputs and outputs of the system
▷ which specific inputs and/or outputs you are interested in investigating (i.e. which you suspect to be unsatisfactory in some way)
▷ the parts of the feedback and control operations which need to be investigated (i.e. which you suspect to be unsatisfactory in some way).

Stage Two

Collect as much information as possible about the way in which the parts of the system you are interested in work. This may involve collecting factual information (e.g. how long or how often something takes to happen) and opinions (people's observations and comments). In particular you need to try to find out:

▷ what happens when things don't run smoothly, or in an emergency
▷ how the system is maintained over different periods of time.

Stage Three

Finally, evaluate the extent to which:

▷ the transformation process is over complicated and wasteful
▷ the limits of the feedback system (i.e. the points at which mechanical adjustments or decisions for action are made) are too high or low
▷ suitable provisions have been made for emergency operation and maintenance
▷ the system can be easily modified to take account of later needs
▷ various changes in the original inputs to the system would improve its performance.

Production Planning Systems (1)

As you develop your design ideas you will need to be thinking carefully about how the various parts of your product would be printed and made most efficiently. You may need to modify your design to make it easier and quicker to reproduce in quantity.

One-off

It takes one hour for someone to make a shirt from a length of cloth. This is called single unit, or **one-off production**.

Batch

If three people work together, sharing the tasks and the manufacturing equipment, they find they can make twenty shirts of an identical design in four hours. They could then quickly switch to making a number of completely different garments to meet market demand. This is known as **batch production**.

Mass

If a number of workers organise themselves and their workplace appropriately then they might easily be able to make ten or more identical shirts an hour, eight hours a day, for weeks on end. This process is called **mass production**.

Continuous

Continuous production is when the production process is set up to make one specific product 24 hours a day, 7 days a week, possibly over periods of many years, such as in some food processing and chemical manufacturing.

Most production processes involve a mixture of these methods. Some parts might need to be individually or batch-produced, while others will be run off continuously.

Different types of manufacturing equipment are needed for the different processes. Some require special-purpose tools, made to suit a particular product. Others require basic machines with parts that can be changed and re-programmed when necessary to make different shapes and forms.

IN YOUR PROJECT

You might decide to use a particular material, or to develop a certain shape or form, simply because it will be easier to manufacture.
Other aspects to consider are:

- how a number of people might work together to make something
- when various stages of the process can be automated.

KEY POINTS

One simple manufacturing technique is the idea of the template – a standard shape which is used as a pattern for cutting identical shaped pieces.

Another thing to look out for is simple batch-production – cutting or assembling a number of components all at once, rather than individually.

Using a mould is another way of producing identical copies of a shape or form.

Production Planning Systems (2)

It is possible to define the step-by-step process by which something is going to be made in very simple terms.

The different stages of production can be grouped together into key areas of manufacture.

When planning a production line, different operations can be coded by using different symbols.

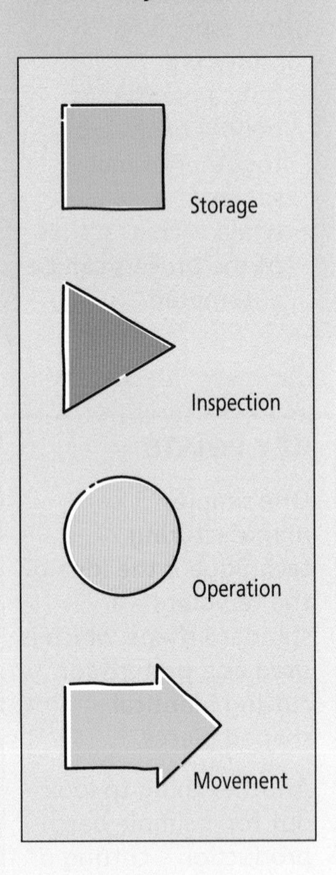

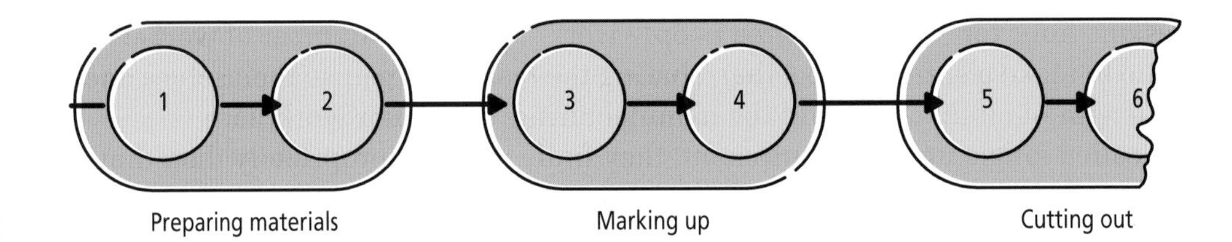

Flow chart grouped into simple stages

Across the stages of production there are likely to be many sub-assemblies. These are groups of component parts being assembled before they are added into the main production line.

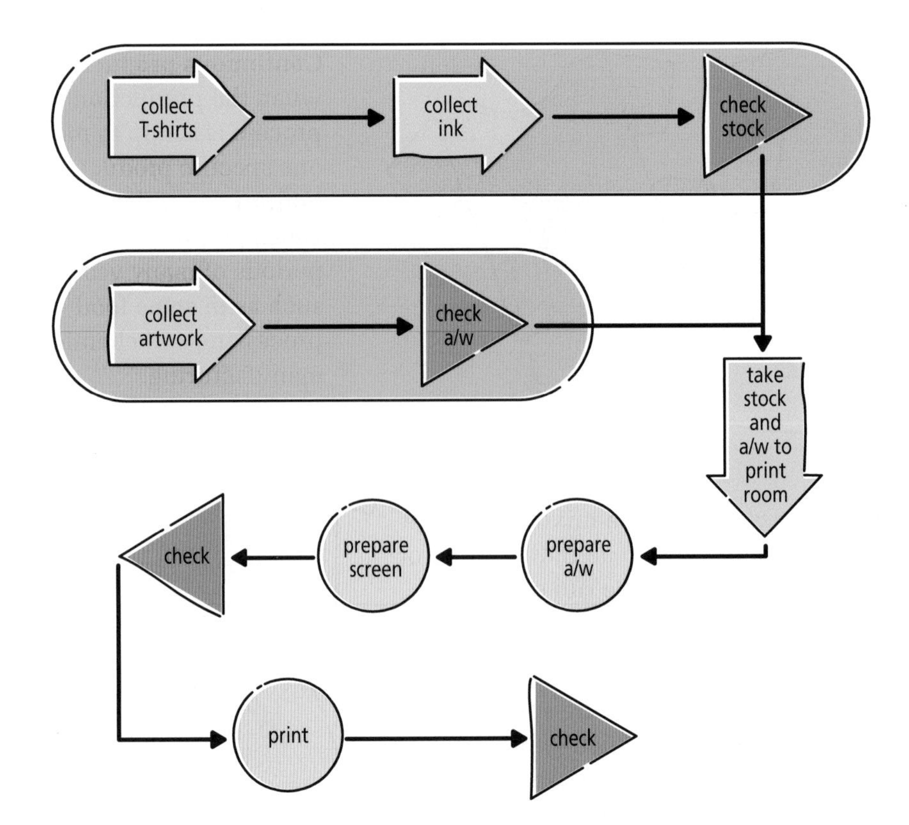

Flow chart for the making of T-shirts.

When all the operations have been identified, the next stage is to plan the layout of the production line.

Sub-assemblies are often made in what are called **manufacturing cells**. These are smaller individual units of machines and operators.

Other ways of organising production processes involve grouping similar machines and/or materials together in one area. This has some advantages, but generally increases the distances which components need to travel.

A conveyor-belt approach is well suited to some types of product, but relies on a steady supply of parts and continuous operation: a delay or breakdown at any one stage can slow down or stop the whole system.

For large single items, such as ships or communication satellites the product needs to be fixed in one place, with the workers and parts organised to arrive at the work at the right time.

In all types of manufacturing production, a complex and accurate production schedule (e.g. a Gantt chart, shown right) is essential to tell the assemblers when to make and assemble all the different components.

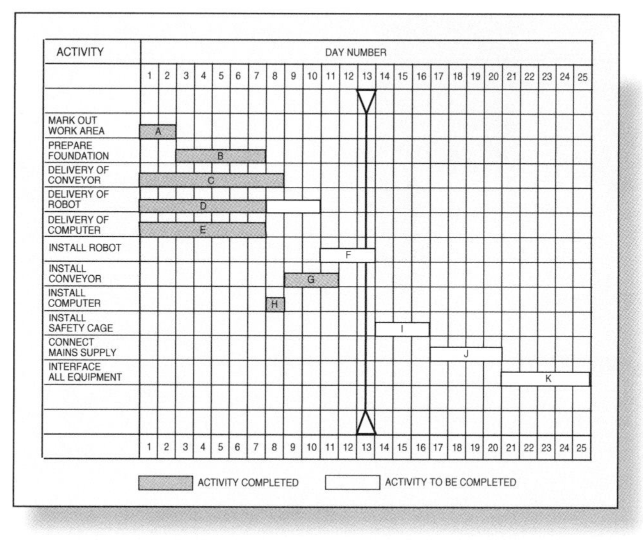

KEY POINTS

Different products require different methods of production. The method chosen usually depends on:

- the number to be made
- the speed of production
- the size and complexity of the product.

IN YOUR PROJECT

As you develop your design ideas you will need to be thinking carefully about how the various components of your product would be manufactured most efficiently.

Just in Time

Manufacturers use 'Critical Path Analysis' and 'Just In Time' methods to produce highly efficient working schedules.

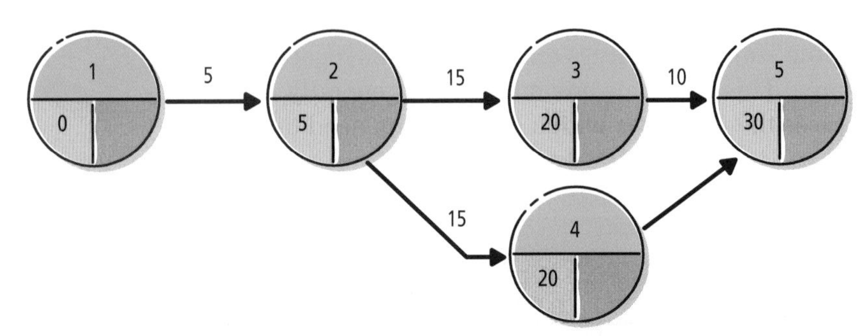

Section of a critical path analysis. The number 1–5 represent stages of the process. The numbers over the arrows represent the time (say in weeks) between stages. The numbers in the left-hand quarters represent the time since the project started.

Critical Path Analysis

Critical path analysis (or CPA) is a technique used for determining the overall schedule of a project. It involves defining the task, identifying its component stages and the order in which things need to be done. Estimates are made of how long each operation is likely to take.

A network path of the essential activities which must be completed before others can take place can then be created. This provides an indication of the minimum amount of time which will be needed to schedule for production.

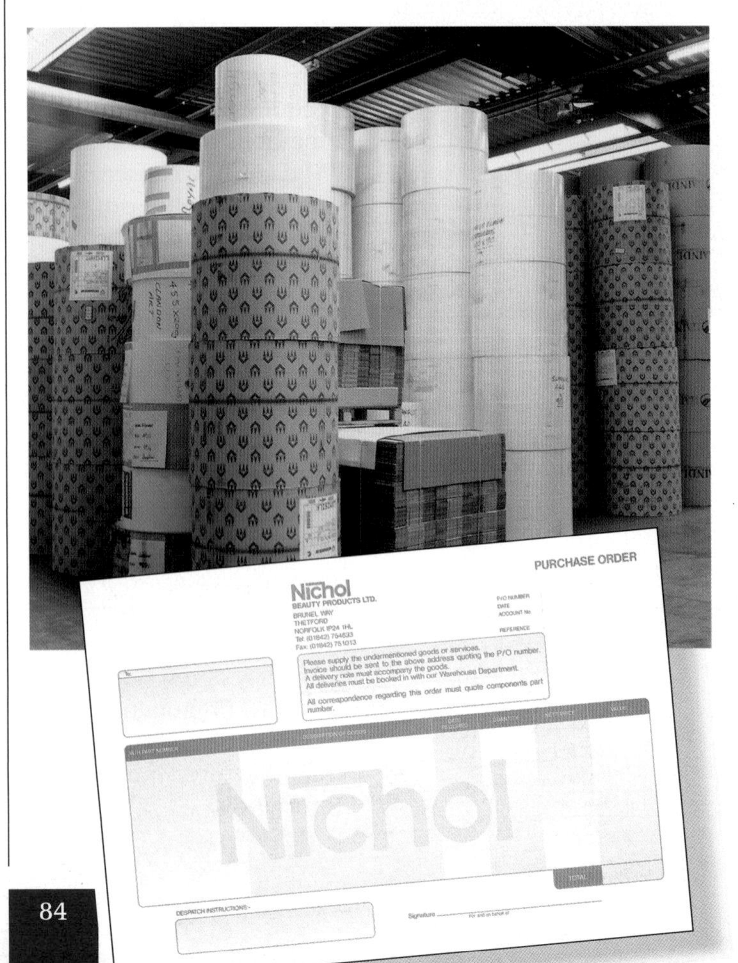

Just in Time

Just In Time is a material and production control system which ensures that materials and components arrive in the factory and at the assembly line at exactly the right time for the product to be made. This helps eliminate excessive use of storage space and the possibility of running out of essential items. To achieve this, the type and layout of the factory, the production run set-up time, work-scheduling and production quality control all need to be considered.

The term can also be used to apply to the efficiency of delivery to a customer, at the right place at precisely the right time.

Factory layouts and production scheduling can now be made considerably quicker by using computer-based CAD and DTP systems.

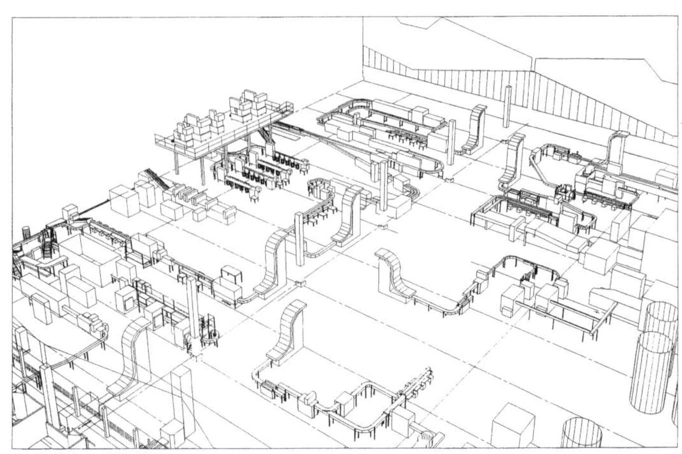

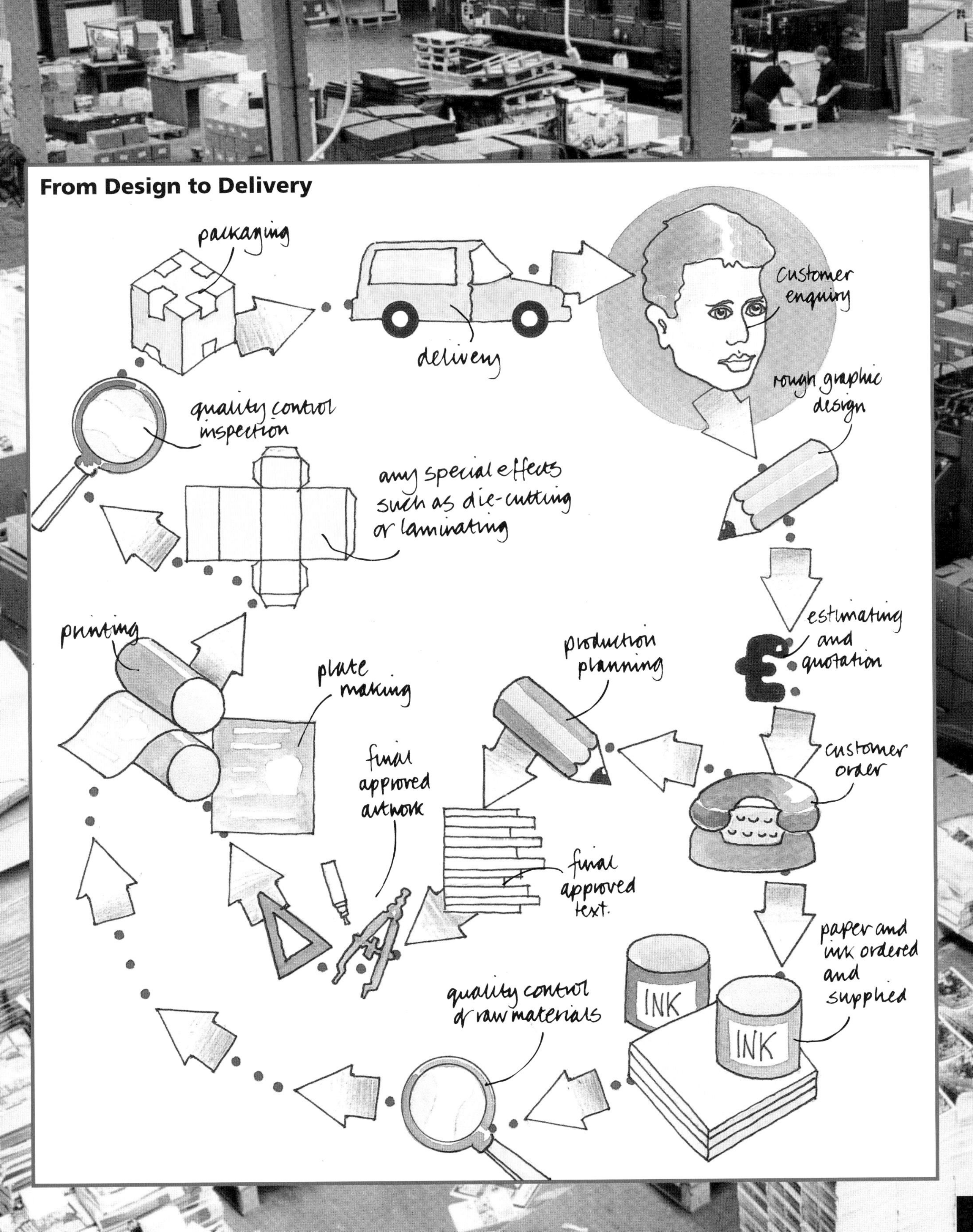
From Design to Delivery
packaging
delivery
customer enquiry
rough graphic design
estimating and quotation
customer order
paper and ink ordered and supplied
INK
INK
quality control of raw materials
final approved text.
final approved artwork
production planning
plate making
printing
any special effects such as die-cutting or laminating
quality control inspection

Quality Counts (1)

Manufacturers need to ensure that all the products they are making are of acceptable quality. A range of techniques have been developed to help check and maintain quality over a long production run.

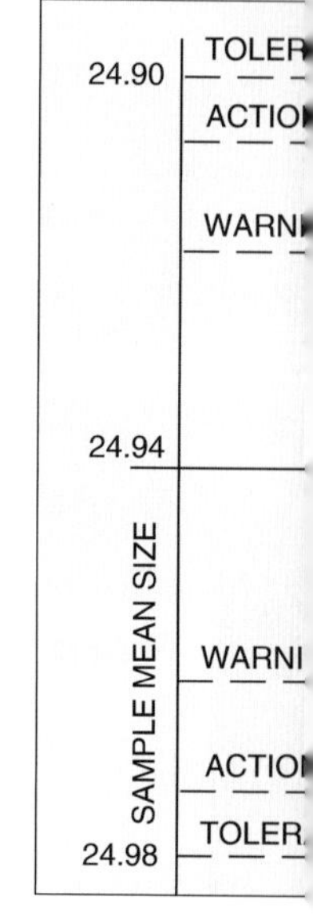

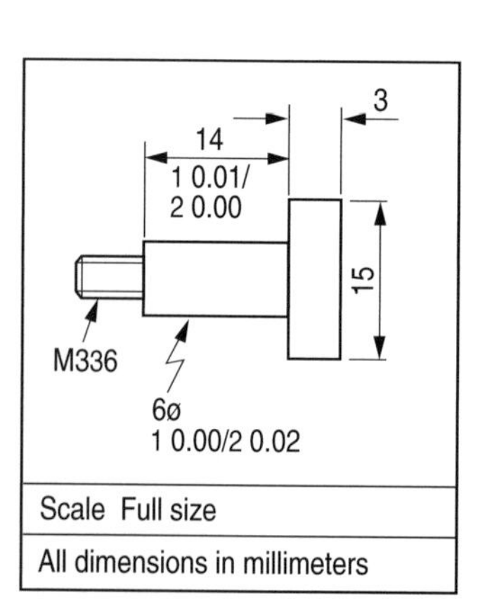

Working to Tolerance

Have you ever tried making something exactly to size? It is unlikely that you succeeded: what you produced would almost certainly have been out by fractions of a millimetre. For you this might not have mattered, but in highly complex products a high degree of accuracy is essential to ensure that important parts fit together exactly. The key question becomes; how accurate does it need to be?

The answer to this question is known as the **tolerance limit** – the acceptable deviation from the ideal size. Tolerance limits are expressed by two numbers: an upper and lower limit. In a simple example a component intended to be 100 mm in length could vary between 99.1 mm and 100.9 mm. The tolerance is the difference between the upper and lower tolerance limits, i.e. 1.8 mm, or +/– 0.9 mm.

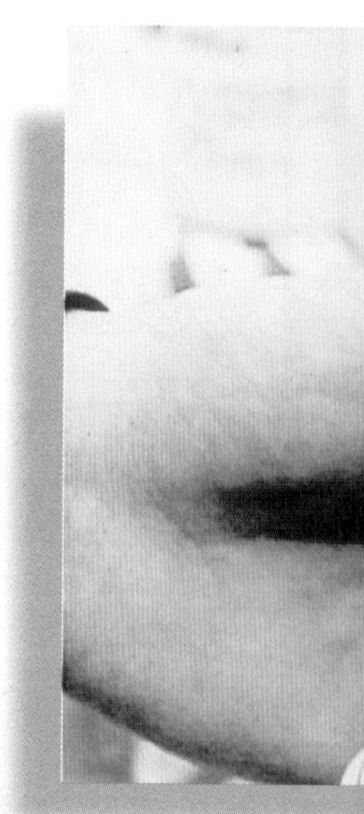

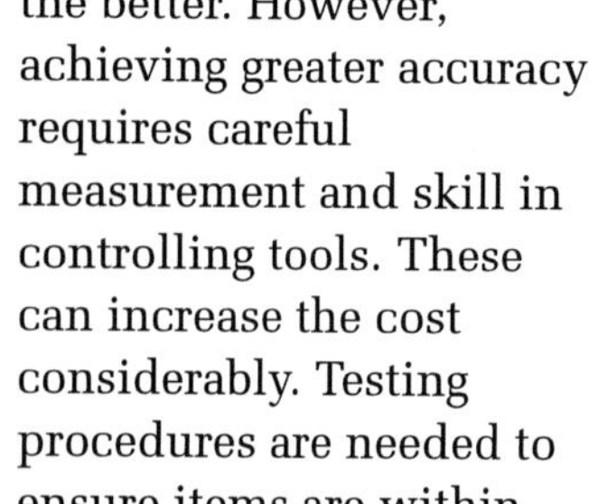

The smaller the tolerance the better. However, achieving greater accuracy requires careful measurement and skill in controlling tools. These can increase the cost considerably. Testing procedures are needed to ensure items are within the stated tolerance limits.

New automated equipment tends to be quicker and more efficient at producing and testing components which are finely toleranced.

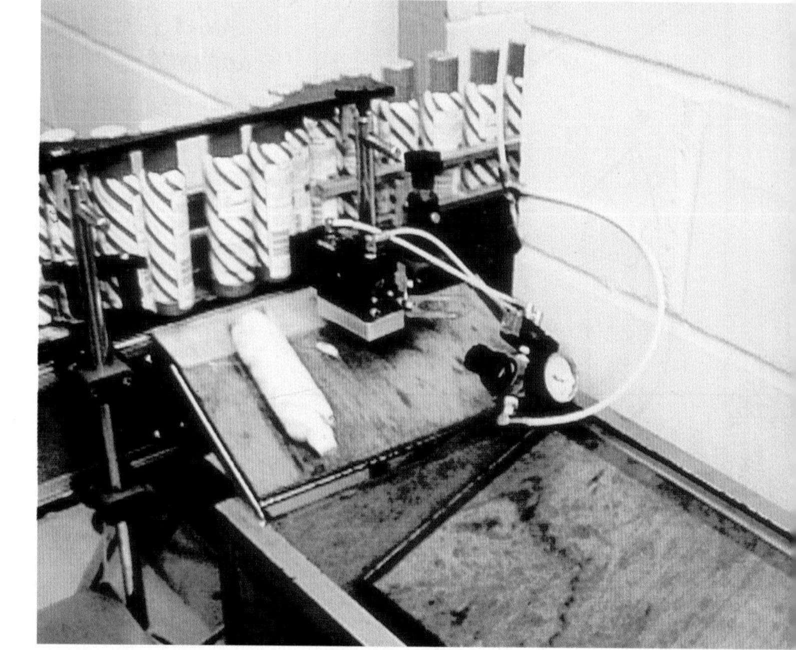

KEY POINTS

- The design specification for a product should include a statement about its tolerance limits.
- Tolerances are important to ensure reliability, which in turn reduces wastage of products during manufacture.
- The smaller the tolerance, the higher the manufacturing cost.

SAMPLE MEAN SIZES PLOTTED

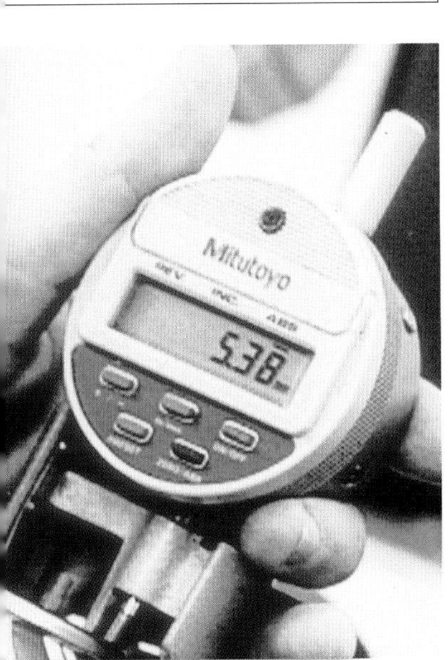

Quality Control

One way to check the quality of products as they are made is to inspect each one to ensure they are all satisfactory.

A more sophisticated approach is to set up a system of **quality control**. This involves inspecting a sample of items at different stages while they are being made, and gathering and analysing records of failure.

A sample of products (say 1 in every 100) are subjected to rigorous tests which identify and record how close the item is to its tolerance limit. If is found to be within acceptable limits, production continues.

By examining the pattern of a series of tests it may be discovered that a particular machine is starting to produce products which are close to being outside the acceptable tolerance limits. When this starts to happen the machine must be adjusted or, if necessary, repaired. If this is not done the machine would soon start to produce items which would be defective. This would be a waste of materials and cause expensive delays in the production line.

The aim of quality control is therefore to achieve 'zero defects' by being able to predict the failure of a machine.

The use of automated testing machines and the electronic gathering and analysis of data lead to higher standards of quality and less wastage.

IN YOUR PROJECT

- How accurately made do the different parts of your design need to be?
- Which components need to fit together most accurately?
- At what stages of manufacture would you recommend that a sample of your product should be measured for accuracy?
- What inspection and measurement tests could be carried out?
- How often should they be done?

KEY POINTS

- Quality control systems help manufacturers reduce wastage and delay in production.
- They do this by predicting failure before it happens

Back to Badges

Imagine you intend to make a range of coloured badges to sell at a local charity or school fair.

First develop a series of design ideas for the badges. Think carefully about how easy or difficult it might be to produce the artwork for each design, and how much the materials for each would cost. A photocopier might speed things up, but will increase the production cost.

Working as a group, choose six designs which you are going to manufacture.

Devise a series of experiments to simulate a range of alternative production processes. For example, how long does it take one person to produce the artwork for one badge and then to manufacture it?

Compare this with how long it takes different numbers of people who have broken down the task into various stages.

- What effect do different tolerances make?
- Where and how could you build quality control checks in?

Try to work out the most efficient way to make different production runs.

Prepare a short formal report on your findings.

Quality Counts [2]

CHECKING PROOFS.

Check type is not broken, missing or illegible.

Check colour bar on each page to ensure colours and tints are consistent in density. This can be done electronically using a densitometer.

Check colours are correctly aligned using the registration marks. The marks should appear black.

Check illustrations for contrast, detail and colour balance.

DESIGN & MAKE IT: GRAPHIC PROD

Any Colour You Like! [2]

Colour Fusion

When small areas of colour are seen together the colours merge and look different. For example a mixture of blue and yellow dots will give the impression of being green. This is known as **colour fusion**. Try looking closely at a large advertisement hoarding or a TV screen to see examples of this.

George Seurat used dots and small brush strokes to create his paintings. His technique is known as Pointillism.

Colour Separation

Modern colour printing uses dots of the three primary colours (yellow, magenta and cyan) to create the full range of colours when mixed together on the page. These are known as **process colours**. Colour separation is described in more detail on page 67.

26

Quality Assurance

What is 'good quality'? This can be a difficult question to answer. A good quality product is not necessarily expensive, or one which lasts for ever. We might say that a product that does what it is supposed to do and is safe is good quality.

For the consumer, a product is of good quality if it is felt to be good value for money. For a manufacturer, quality is also about making such products in the most efficient way, and therefore for the most economical price.

Quality Assurance is the overall approach which ensures high standards of quality throughout a company. It includes the development and monitoring of standards, procedures, documentation and communication across the whole company. Usually a quality manual is produced which contains all the relevant information to guide staff.

Quality does not just happen: it has to be planned for, and managed. **Total Quality Management** (or TQM) is an approach to management which aims to maximise the human and physical resources of an organisation in the most cost-effective way to meet the needs and expectations of the customer and the community.

IN YOUR PROJECT

- ▶ Which British Standards would apply to to the production of your design?

KEY POINTS

- Quality control production checks form a specific part of a broader programme of quality assurance across a company.
- British Standards exist to define acceptable standards of production

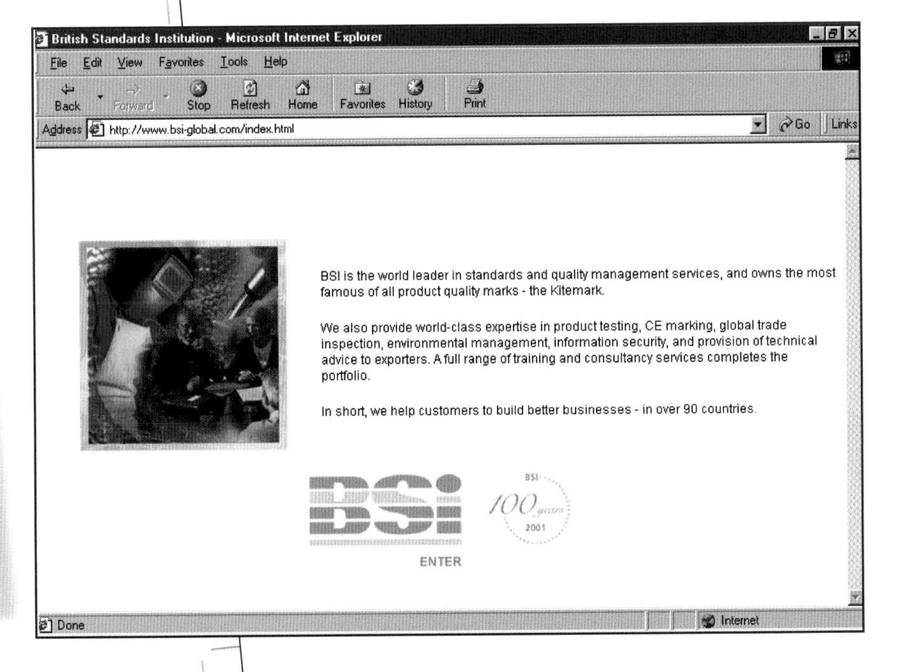

British Standards

The British Standards Institute was the first national standards body in the world. Its main purpose is to draw up voluntary standards to be observed. They produce documents which clarify the essential technical requirements for a product, material or process to be fit for its purpose.

There are a range of over 10 000 British Standards for almost every industry from food to building construction and textiles to toys. They cover all aspects of production, from materials to management.

Certification that a product manufacturing or management process conforms to a stated British Standard provides an assurance that an acceptable quality can be expected. This greatly reduces the risk of someone buying goods and services which could be defective in some way.

Making and Manufacturing

Mix and Match

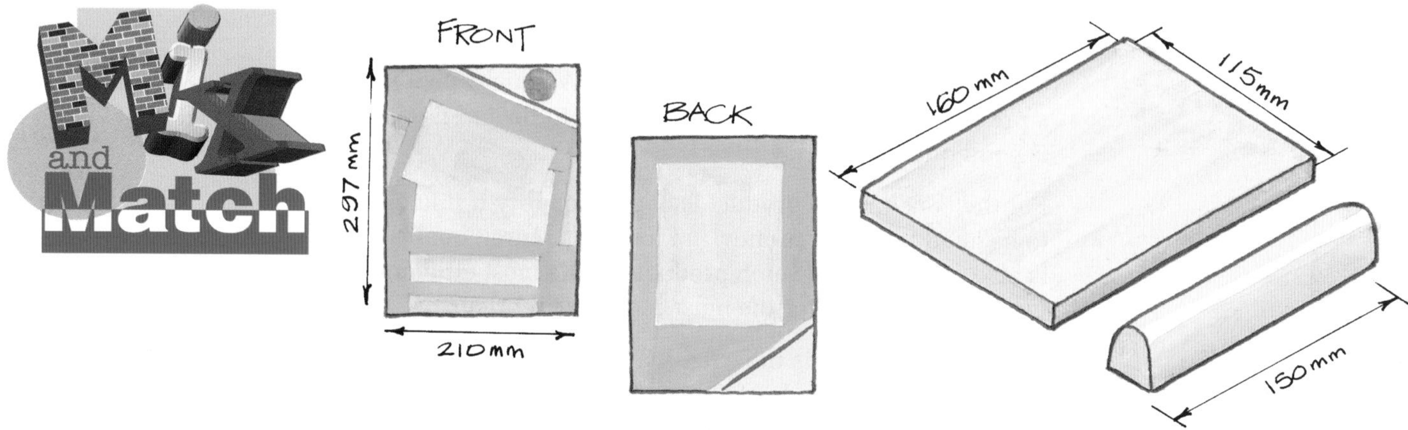

Planning the Making

Make sure you have a final drawing of your package design. You may need more than one to show different aspects of the design. The drawings should clearly show the sizes and shapes of the packaging materials, together with the graphics.

Prepare a list of all the different materials and making stages you will need. Estimate how long each main stage will take. Graphic work should be applied to paper and card before it is folded up.

What operations could be going on at the same time?

Draw a flow chart to illustrate what you will need to do and when.

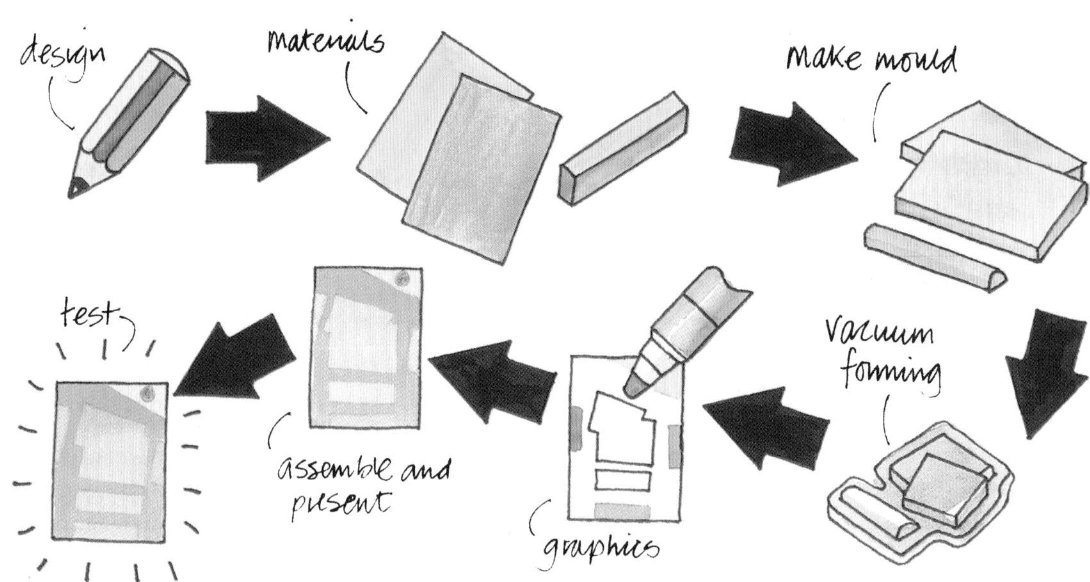

Starting the Report

While you are getting on with the making you could begin preparing your manufacturing and costing report.

If you haven't already done so, you should read and refer to pages 22 to 23 and 70 to 89.

- ▶ What type of paper or card will be used?
- ▶ What printing process would be best?
- ▶ What finishing methods will be needed?
- ▶ How much is it likely to cost to print 500, 1000 and 5000 items?

▷ What tolerances will you be working to? How and when will you be checking that your making is of sufficient accuracy?

▷ What specific safety precautions will you need to observe during the making process?

Testing and Final Evaluation

Final Testing

Look back at your design specification. How well does your solution meet the requirements of the specification?

- Does the package contain the items you stated it should?
- Is the size within the minimum and maximum limits you set?
- Are the materials strong enough?
- Can the items be easily removed without damaging the package?
- Is all the required information on the outside?
- Is the package easy to display and eye catching?

You will need to plan and do some tests to check some of these points.

For example, you could put the package on display in school, or ideally in a local shop, and take photographs of it. Then complete a small survey:

- Do people notice the package easily?
- Can they identify what is in the package?
- Would they buy it, and for how much?
- How easy do they find it to open the package without destroying it?

You could also test the strength of the package by dropping it from a height to see if it breaks, or damages the contents. (A teacher must supervise you doing this to make sure there is no possibility of anyone getting hurt.)

Final Evaluation

Write your final evaluation. Remember to mention good and bad points about the process you used, and how well what you have designed works. Make specific references to pages in you design folder and to the results of the final tests you undertook.

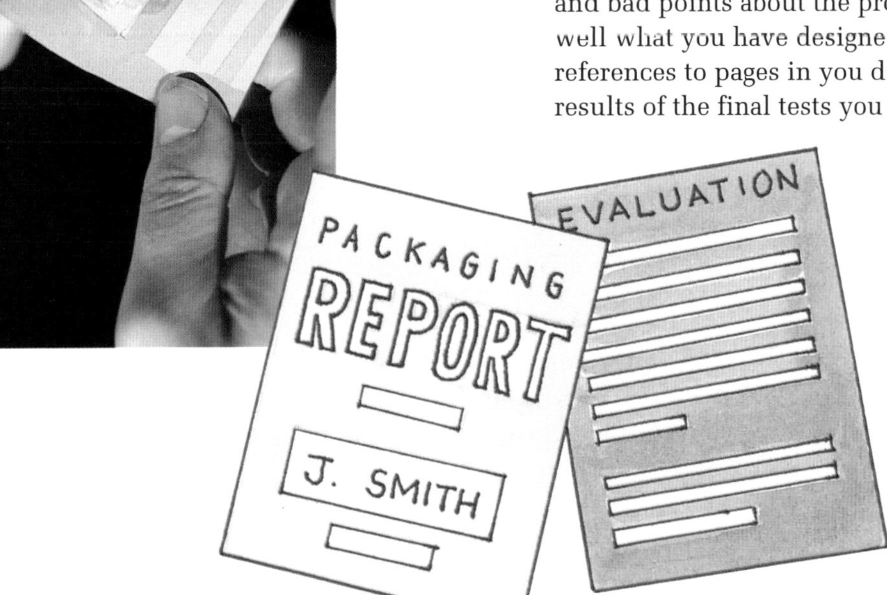

Remember to complete and hand in your manufacturing and costing report. If possible it should be typed, or better still, word processed.

Examination Questions

Before spending about two hours answering the following questions you will need to do some preliminary research into packaging.

To complete the paper you will need some plain A4 and A3 paper, basic drawing equipment, and colouring materials. You are reminded of the need for good English and clear presentation in your answers!

The packaging of products from small sweets to large items of furniture is a multi-million pound industry.

Good packaging helps protect its contents and sell the product, as well as being environmentally friendly and cheap to produce.

Visit a supermarket or 'Do it Yourself' store and study the ways in which small electrical components are packaged. In particular look out for:

- the different methods used to package a number of small items together
- the use of recyclable and environmentally friendly materials.

1. This question is about Evaluation. *(Total 6 marks).* See pages 58-59.

Study the drawing below of a package for holding four domestic electrical fuses.

Describe three aspects of the package that would need further development. *(6 marks)*

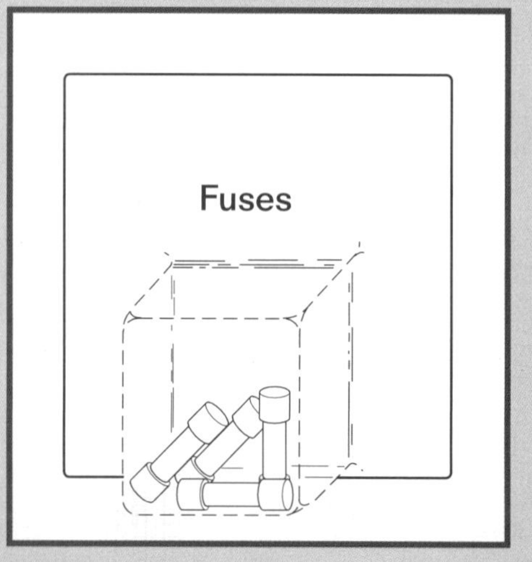

2. This question is about Designing and Making. *(Total 35 marks).* See pages 60-71.

A supermarket called 'EZ-BUY' is to introduce its own range of batteries suitable for small electrical appliances.

You have been asked to design a special promotional package holding four AA size batteries.

EZ-BUY has provided the following specification for the package:

- the supermarket name EZ-BUY and the battery size code AA must be clearly visible
- the package must made from card only
- the batteries must be visible when on display
- the package must hold the four batteries securely
- the package must allow for display from a point of sale stand.

On A3 paper use sketches and notes to produce ideas for:

(i) a net (surface development) and

(ii) a three colour graphic design for a package.

Ensure you provide the following:

a) a range of ideas for the net (surface development) with notes *(8 marks)*

b) an accurate net (surface development) of the best package drawn with instruments *(6 marks)*

c) a range of ideas for the graphic design with notes *(8 marks)*

d) a 3D coloured drawing of the assembled package, including the final graphics. *(13 marks)*

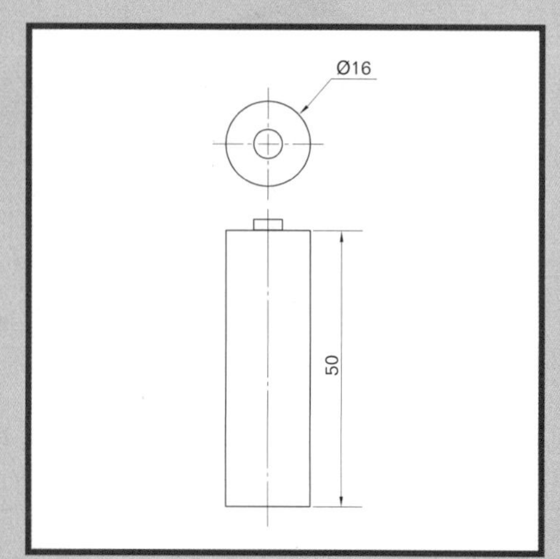

3. This question is about Processes and Manufacture.
(Total 21 marks). See pages 81-87.

a) EZ-BUY wish to see an example of your package and design. List the main stages, in the correct order, involved in making one package, by each stage. Name a piece of equipment you would use. *(10 marks)*

b) Produce a sequential drawing of the manufacture of the example. *(5 marks)*

c) When your package goes into commercial production explain how 'Just In Time' can benefit the production of the complete product. *(2 marks)*

d) What is 'Quality Assurance'? Briefly explain how and why Quality Control is applied to your package. *(4 marks)*

4. This question is about Systems and Control. *(Total 12 marks).*
See pages 78-79.

a) Name the three components of any system. *(3 marks)*

b) Give an example related to the production of your package. *(3 marks)*

c) Explain why registration marks and colour bars are used in the printing of your package and how 'feedback' is used. *(6 marks)*

5. This question is about Materials and components.
(Total 14 marks). See pages 62-67.

a) Name two packaging materials, one a block type and one sheet type. State a typical use for each. *(4 marks)*

b) Vacuum forming is a method of making some packages. Name:

(i) a material that is used for the 'mould' and
(ii) a material that is used to form a container. Explain why these are suitable materials. *(4 marks)*

c) Describe and explain three ways in which designers can make packages more environmentally friendly. Give a specific example for each. *(6 marks)*

6. This question is about Printing methods. *(Total 30 marks).*
See pages 72-77.

a) Name the four main types of commercial methods of printing. *(4 marks)*

b) Match the following products to the most appropriate commercial printing method, and give a reason for its appropriateness:
(i) a newspaper
(ii) a poster for a school play
(iii) postage stamps
(iv) business cards
(v) a supermarket plastic carrier bag. *(10 marks)*

c) Name three special printing effects can be applied to the surface of a package. Comment on their likely effect on the cost of the package. *(6 marks)*

d) What is 'spot varnishing' and 'foil blocking'? Name a product which would be enhanced by the use of each. *(6 marks)*

e) To promote the product a personalised T-shirt is available. Briefly explain how a computer generated design can be put on a garment. *(4 marks)*

7. This question is about Environmental issues.
(Total 7 marks). See pages 68-69.

a) The demands of modern society places great demand on the natural environment. Give two consequences of the over packaging of products. *(2 marks)*

b) Sketch a symbol which shows that a product has been designed with the intention of being recycled. *(2 marks)*

c) Briefly explain why food packages are not made from recycled materials. *(3 marks)*

Total marks = 125

Project Three: Introduction

Presentation Matters (page 121)

3D Computer-aided Design (page 118)

Product Rendering and Modelling (pages 112 and 116)

Drawing in Three Dimensions (page 108)

Read the following newspaper article:

High fashion is not usually associated with high tech. Computers, mobile phones and pagers generally come in tasteful shades of grey and black, with all the style and grace of an engineered brick.

This is beginning to change, however. One company has started selling two-tone mobile phones, available in 11 colour combinations. And a PC manufacturer's products come with a choice of front panels, enabling users to change the colour of the PC to suit the decorations in their home.

A high-street fashion chain has brought a sense of style and colour to pager design, aimed at young night-clubbing teenagers. The unit displays figures, not letters, so messages have to be sent in code. In the US children have developed a whole code-book. 0 means 'no' and 1 means 'yes', obviously enough. 180 means 'there has been a major change of plans, and 360 means 'back to original plans'. 13 is 'bad luck', and 1402, as in Valentine's day, is 'I love you'. Once the pager is purchased there are no further subscription charges. The system is funded through a flat-rate charge for calling the pager.

ACTIVITY

- In what ways are electronics manufacturing companies trying to make their products more attractive and desirable?
- What particular sort of people are their new products aimed at?
- What number-based codes can you invent to represent pager messages?

'As you know, In Touch Telecommunications have commissioned us to develop design ideas for a new pager communication system they are developing.

We now have details of the minimum size the electronic components can be housed in, and a list of the various controls and displays which are needed.

We need to suggest what sort of people the pager will be aimed at, how it will be used and what it will look like.

In Touch would also like to be advised on how a high quality, reliable product can be manufactured at the same time as keeping production costs to a minimum.'

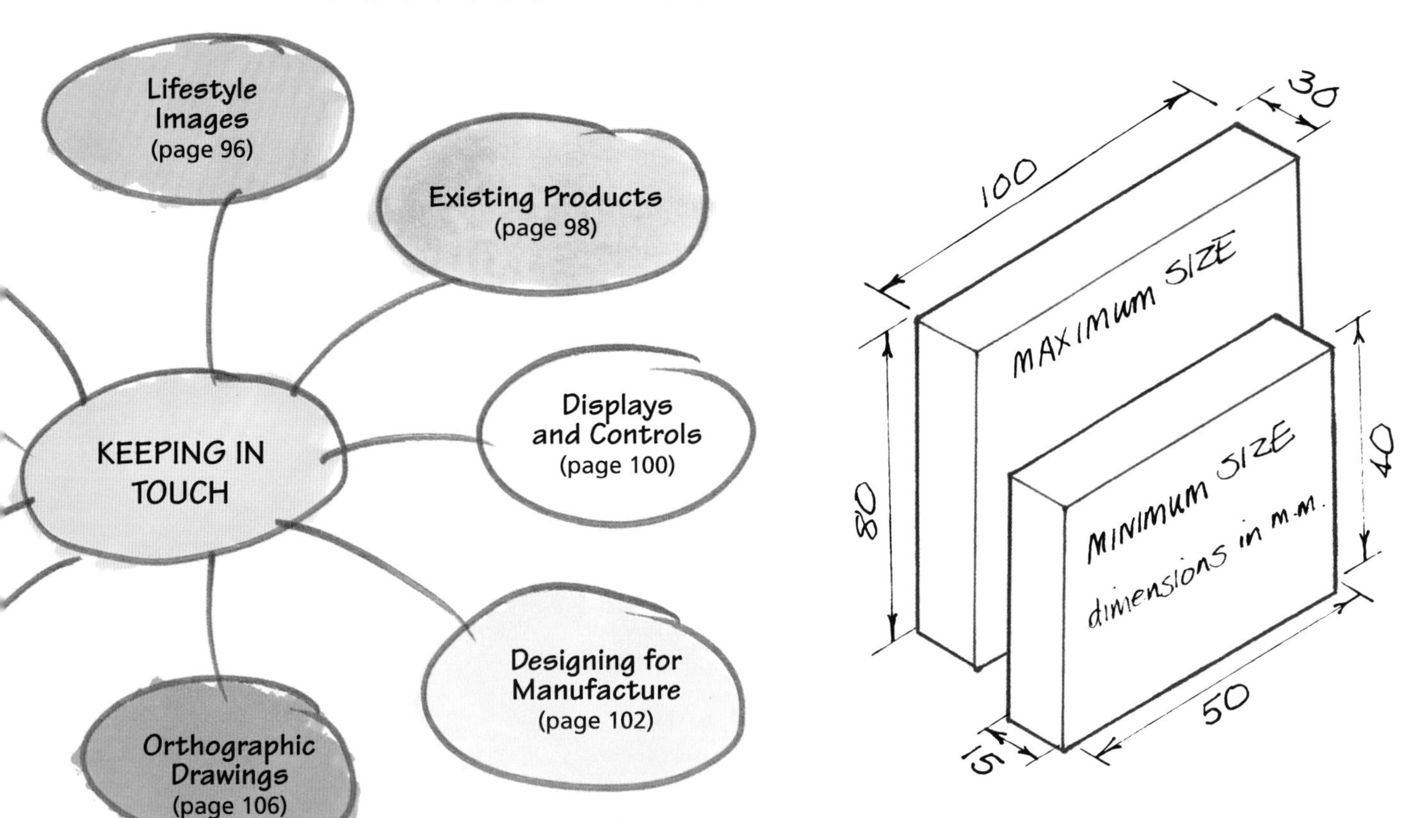

Summary of Project Requirements

In Touch are looking for suggestions for:

- ▷ the target market for the pager, and its characteristics
- ▷ the graphic identity of the product, expressed through the shape, colours and graphic details of the casing
- ▷ how the device will be held and carried, the size and position of the displays and controls.

What ideas can you come up with?

CLARIFYING THE TASK

You have been given a lot of information about a new product which has already been developed quite a long way.

- ▶ What exactly are you being asked to design?
- ▶ What presentation boards must you prepare?
- ▶ What other presentation items might you include?

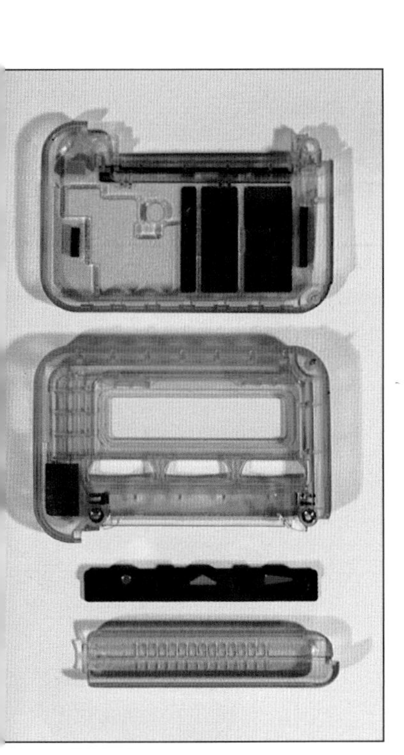

Presenting to the Client

You are asked to prepare materials for a presentation session to In Touch Telecommunications. These will need to include a display panel for each of the following:

- ▷ Information about the lifestyle of the target audience you choose for the product.
- ▷ A coloured 3D rendering of the pager, highlighting the design features and the layout of the display and control devices.
- ▷ The product drawn in dimensioned orthographic.

In addition you might also include:

- ▷ A display panel showing the product in use (a drawing or photograph of a product model).
- ▷ A display panel showing a cut-away or exploded view of the product.
- ▷ A short illustrated technical report on the manufacturing issues for the production of the pager.

AN ALTERNATIVE TASK

In Touch are also looking for ideas for mobile phones with text messaging systems incorporated. Instead of designing a pager, you could develop ideas for a new mobile phone with Internet access, aimed at a specific market.

Lifestyle Images

People have a wide variety of individual needs and wants. However certain groups of people tend to think and act in similar and sometimes surprising ways. New products can therefore be targeted to appeal to particular markets.

Lifestyles

People are all individuals with their own specific likes and dislikes. We can choose what we want to do and where we want to go. But how unique are we?

Many adults who drive particular types of car enjoy similar sorts of leisure activities. People who regularly eat convenience foods share a style of life that means they also buy similar magazines.

Market-research organisations have come up with a range of ways of classifying lifestyle groups.

For example:

▷ 'Homemakers' are reliable, self-sufficient people, concerned with their local community.
▷ 'Achievers' are hard-working, highly competitive, ambitious individuals who enjoy using, and being seen to be using, the latest technological products.

Products and services can be aimed at particular life-style groups, and their design and advertising can be developed to make them appropriate to the environment and situations those people might expect to be in.

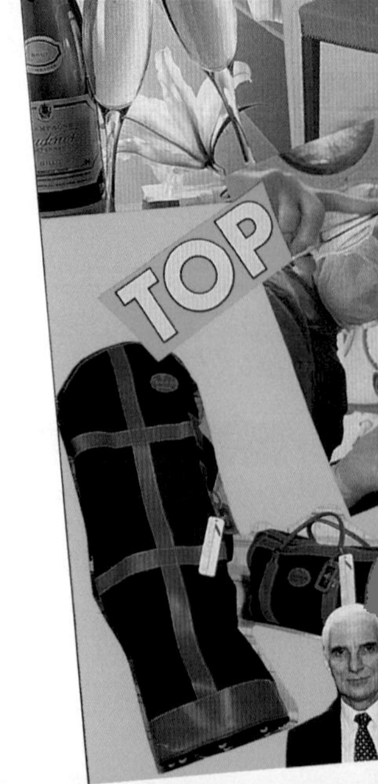

Finding The Gap In The Market

We buy things to satisfy our desires. We don't just buy a personal stereo, for example. We want one in a particular price range which will have a specific range of functions, and looks the way we want it to look to reflect our lifestyle.

Manufacturers produce a range of models to satisfy different markets. Companies are keen to spot a gap in the market, i.e. a product model or variation which is not well supplied by other manufacturers.

Your Target Market

You will need to choose a target market for the new pager. At present there are plenty of gaps in the market for pagers aimed at particular situations and circumstances. For example, yours could be for one of the following groups:

- ▷ doctors, lawyers, pilots, headteachers
- ▷ designers, journalists, advertisers
- ▷ college students
- ▷ social services, health-care providers
- ▷ musicians
- ▷ sports professionals
- ▷ some other type of people.

The one you choose will strongly influence the overall appearance of the design you develop.

■ ACTIVITY

Prepare an image board based on your chosen target audience for the product. Start by making a list of:

- ▷ the activities they typically do (sports, hobbies, holidays, etc.)
- ▷ the places they might go to (shopping centres, seaside, foreign travel, etc.)
- ▷ the things they buy (cars, clothes, food, etc.)

Then look through magazines and brochures to find images of these particular activities, places and products.

Think about colours and textures. Look for appropriate words used as titles – these might be adjectives, e.g. 'fun-loving', or nouns, e.g. 'Paris', 'Skoda', 'Lager'. In particular look out for examples of people communicating through speech, gesture, word, picture, telephone, etc.

If you can't find enough images or words, you may have to do some as coloured drawings/lettering of your own.

As you develop your image board, remember the following:

- ▷ start in centre and work outwards – avoid blank spaces in the centre
- ▷ experiment by allowing some images to overlap or form simple patterns
- ▷ try including some blank sheets of coloured paper as part of the background
- ▷ make sure images are well balanced across the board
- ▷ try placing some images at an angle, but not too steep
- ▷ cut some images out completely from their background
- ▷ think carefully about the size and placing of lettering.

Evaluating Existing Products

Market researchers and designers often test and evaluate existing products and services in detail. They do this in order to identify ways in which future designs might be improved in terms of manufacture, performance and sales potential.

Search the web to find out more about the history of telephones. Try:
www.bt.co.uk
www.sciencemuseum.co.uk
www.designmuseum.org

■ ACTIVITY

Undertake an evaluation of an existing telephone (wall-mounted or portable), or some other hand-held electronic device, such as a personal stereo, portable radio, camera, electronic game, etc. Don't attempt to undo the casing or to remove any of the component parts.

- ▷ What particular features does the device have?
- ▷ What market is it aimed at (i.e. person and situation)?

In particular look at its colours, textures and detailed shapes of the casing. Also look at the arrangement of displays and controls (see page 100).

- ▷ How well do the size and positions of its displays and controls relate to average hand sizes while the device is being operated and carried?
- ▷ What materials is it made from?
- ▷ How do you think it might have been manufactured?
- ▷ How does it compare with similar devices produced by other manufacturers?

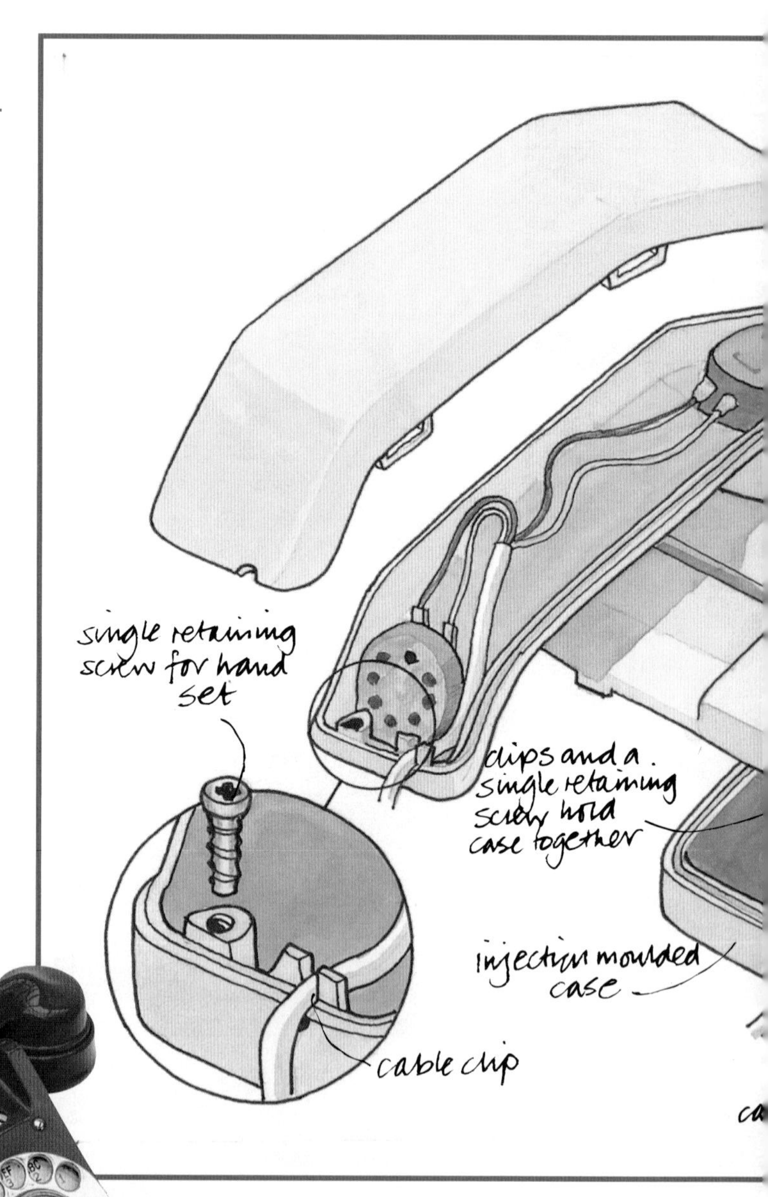

Human Factors

Devise a comfort/appearance rating test. Using a 1 to 10 scale ask at least ten people to score questions such as:

- ▷ How easy do you find it to...?
- ▷ How well does the product fit in with...?
- ▷ How comfortable is it to hold...?

Plot the results for each question in an appropriate graphical form, such as a bar chart, with the rating on the horizontal axis and the number of people on the vertical axis.

Design History

Find out what you can about the history of the design development of the device.

▷ When was it first introduced in its present form?
▷ How has its shape, form and features evolved?

Make a statement about:

▷ the gap in the market the device is aiming to satisfy
▷ the sort of people it is aimed at
▷ where it is in its product lifecycle.

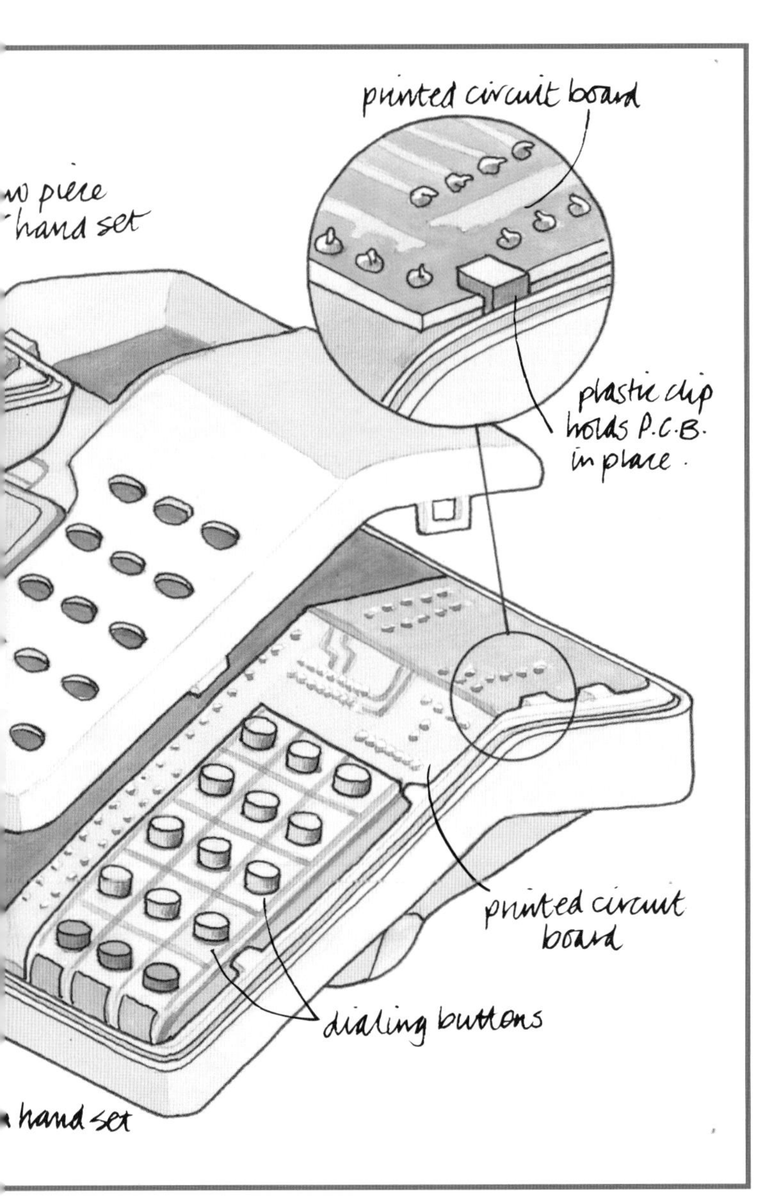

The Product Lifecycle

Products are not intended to last for ever, and have a lifecycle which usually lasts over a period of years.

When a new product is first launched sales are likely to be slow. Profits from sales are unlikely to provide a return on the development, mass manufacture and initial promotion costs. In the second phase of a product's lifecycle demand picks up as the product becomes known and accepted. Everyone starts to want one, and sales repay the initial investment and show a healthy profit.

In the third phase the product will be well established and selling well with minimum promotion, but also starting to compete with newer models being produced by other companies. These may well be cheaper or have a better performance specification.

In the final phase, sales will drop off. This can be rapid as rival products now in their second and third phases begin to dominate the market. The cost of production may exceed the potential profit, so the model is withdrawn. Sometimes manufacturers re-launch products with minor modifications to help extend the sales period.

Most design is evolutionary. New products are rarely completely new, but an adaptation of an existing design. They might incorporate new features, say a more aerodynamic shape, or use a number of different materials and components to produce a better-looking, better-working model which is cheaper to manufacture.

Displays and Controls

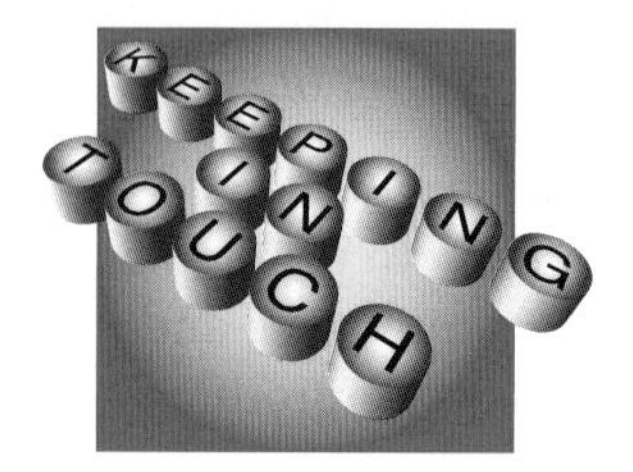

Effective controls and displays are needed if products are to work properly. Displays provide the operator with important information about the state of the machine. Control devices enable them to make the necessary adjustments.

A very important part of the job of the designer is to ensure that products are easy to understand and use.

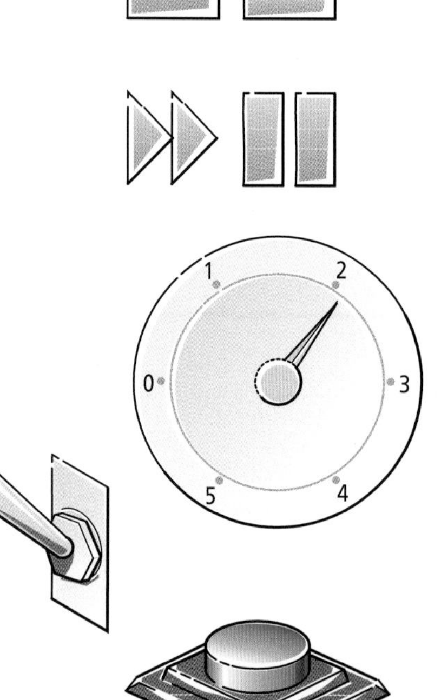

Most mechanical and electronic products work by means of a series of events happening in sequence. The movement of a number of mechanical devices and the flow of current in electronic circuits need to be controlled in a variety of ways – switching them on or off, making them work faster or slower, for example.

What a display or control looks and feels like, how it is operated and where it is placed can all make a great deal of difference. Often it's not so much the basic displays and controls used everyday that can cause problems, but the ones that are used infrequently and sometimes have to be set in a hurry or in an emergency.

At the Interface

The displays and controls of a product are often called the **interface** – the point at which the person and the machine interact.

Displays (such as lights, graphic symbols next to switches and dials, and sometimes synthetic voices) pass information to people about the current state of a machine or device. Is it on? How hot is it? What setting is it on?

Controls (such as switches, dials and touch-sensitive buttons) enable people to alter the state of the machine, e.g. switching it off, cooling it down, changing the setting.

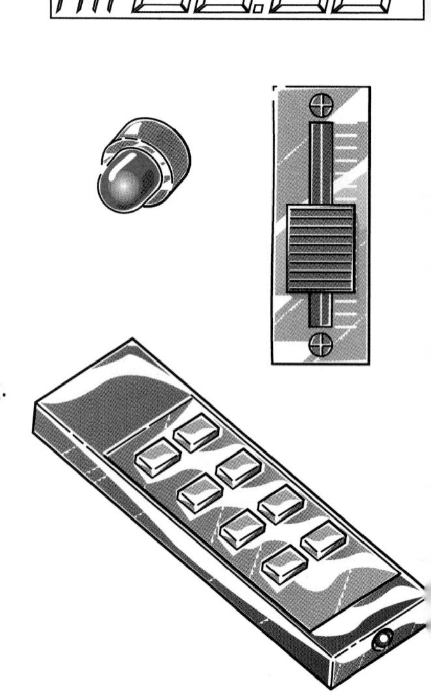

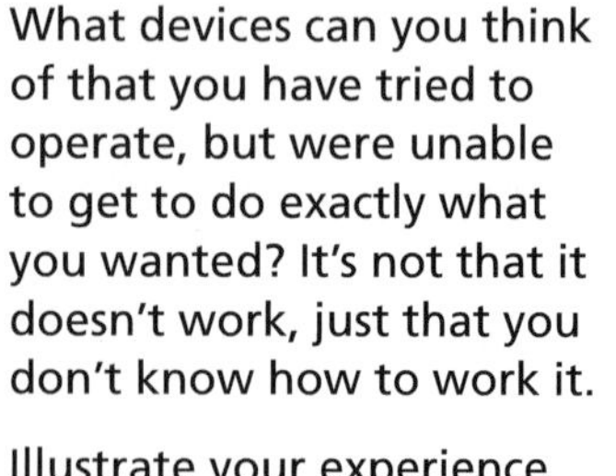

■ ACTIVITY

What devices can you think of that you have tried to operate, but were unable to get to do exactly what you wanted? It's not that it doesn't work, just that you don't know how to work it.

Illustrate your experience like the cartoon on the right.

Gear stick

Hand brake

Oil warning light

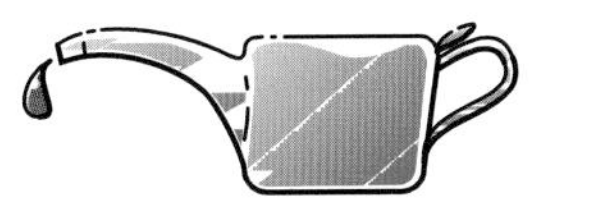

Hand brake engaged

Control Panel Layout

Different controls and displays are used at different times while the product is being used. Each one can be identified as being either:

- **on-line**, i.e. those which are used while the device is in use to monitor and control the system
- **off-line**, i.e. those which need to be used immediately before the product is used, or when it is due to be finished with.
- **danger**, i.e. those which warn that something is wrong
- **maintenance**, i.e. those which are used periodically to ensure everything is working properly.

These displays and controls are often grouped together. For example, all the on-line displays and controls might be placed where they are easiest to get at. Sometimes the first off-line control, used to switch the device on, is on the far left hand-hand side. Warnings displays are often placed with the on-line controls so that they will not be missed while the device is being used.

■ ACTIVITY

What types of displays and controls do drivers of motor vehicles need? Generate a list of as many as you can think of. Group your list together under the headings in bold on the right.

IN YOUR PROJECT

Think carefully about how someone will use the displays and controls you are designing:

- will it be easy to turn on and off, and to adjust?
- will they assist safe operation?
- would a visual or aural display be most effective?
- how might new electronic, computer-controlled display and control systems be used?

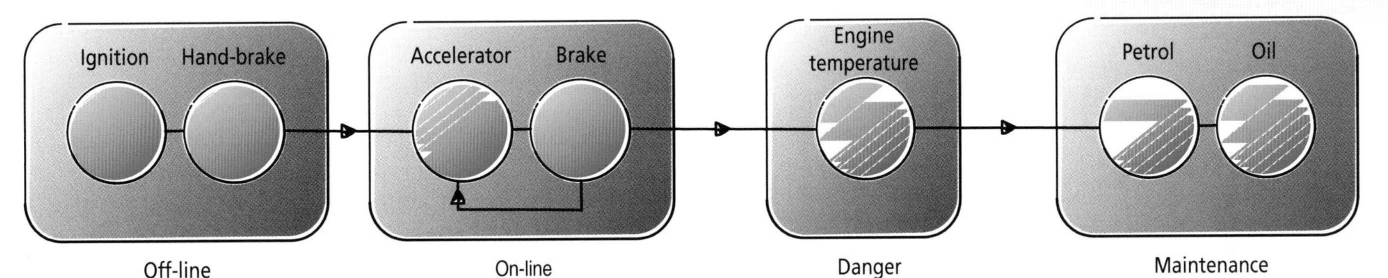

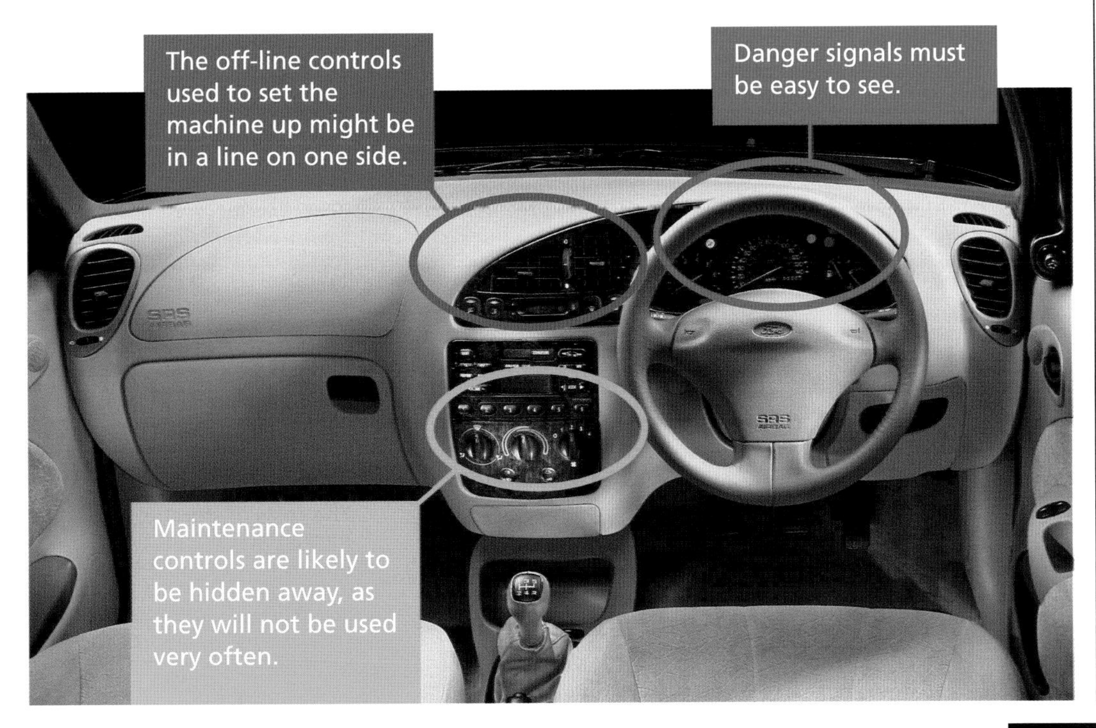

■ ACTIVITY

- Look at the illustrations on this page and page 100. Which are displays and which are controls? Which, if any, are both?
- How have graphic symbols been used to help display the state of the controls?
- Select a hand-held device and evaluate the type and arrangement of its displays controls

Designing for Manufacture (1)

There are many different things which need to be taken into account when designing something which is suitable for manufacture. These include planning for the available production resource, the expected life of the products and the most cost-effective method of manufacture.

"We needed a design that was easier, and cost effective to manufacture in large numbers, used fewer parts, was attractive and user-friendly"

What Do Designers Do?

One of the key skills of a designer is taking an existing scientific principle, such a simple motor, working out how it can be held in a safe as casing, and how the various elements can be assembled quickly and easily. At the same time, the product which has been made must work well and be needed and wanted by enough people to make the whole operation financially viable.

Designers often have to design within considerable constraints, however. They rarely have a free choice of materials, components and production processes, and often have to work with what a manufacturer already has.

Further limitations may be imposed by things like the maximum size a machine can mould, an existing stock of ready-made electronic components and the existing skills of the workforce.

Some products are created with the prime intention of utilising materials and equipment which are being under-used during the decline of a particular product.

Design Failures

Sometimes products fail to sell. Maybe no-one wanted it because there was a better or cheaper alternative. Or perhaps it quickly became known that it didn't work well, or was unreliable. Maybe not enough money was invested in promotion, with the result that not enough people knew it was available. Possibly manufacturing costs proved to be much higher than expected, with the result that no profit was made.

Finding Fault

Much work goes into improving the design of a product. Companies often test their products using a variety of information and data-gathering techniques. Its performance is then compared with their competitors' products. Sometimes a fault may not be so much with the original design, but in the quality of manufacture.

In recent years a major concern for many companies has been in minimising environmental pollution in the manufacture and use of its products. Methods of manufacture may be analysed to determine potential modifications.

Reducing Costs

Frequently the emphasis is on reducing production costs. Some elements will be fixed costs, while others are known as variable costs. **Fixed costs** are those incurred in setting up an assembly line, such as machines, tools and factory space.

Variable costs are likely to change according to the number of products being made, and cover things such as raw materials, energy, staff wages, insurance, maintenance, etc.

The costs of storage, packaging, distribution and selling all need to be considered too. VAT is another element which adds to the final selling price of the product.

The actual manufacturing cost of a product in terms of its materials and labour will vary according to the particular item. Often it only represents some 5 to 10% of the final selling price.

The profit made by the manufacturer, or investment organisation, will also vary depending on the product. Typically it ranges from just a few percent for high-volume, rapid turn-over goods, to up to 50% or more for exclusive, high-quality, hand-finished items.

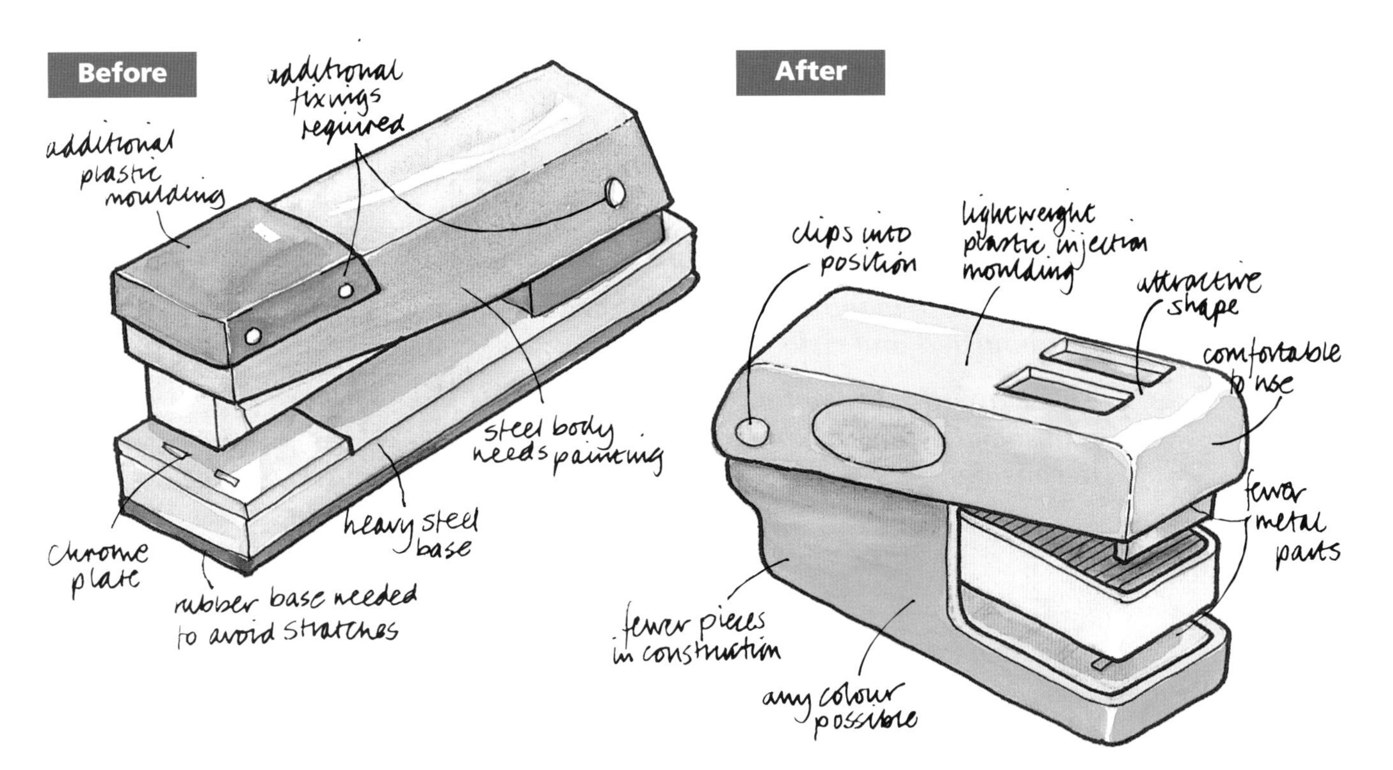

Design for Quantity

Products need to be designed to be easy to make. Some items may prove to be very difficult to make in quantity however. This might be because of:

- their shape
- the way the components are arranged
- the materials required.

Different manufacturing processes and materials can be used according to the numbers to be produced.

The rate of production is an important factor too. Producing 10 000 units by the end of the week in order to satisfy demand will need to be approached in a different way to making the same number over a twelve-month period. There are also important implications if production needs to be organised in batches, for example, 5000 this month and another 5000 in a year's time.

Design for Maintenance

Designers need to consider how often a new product will need to be maintained during its usage, and take this into account while developing ideas.

They will also have to think about how easy it needs to be to undertake the maintenance work. If a component needs cleaning, adjusting or replacing often by the user (e.g. replacing a battery) it must be quick and easy to do. Other maintenance might need to be done by trained specialists however, and providing easy access might result in damage if the user tries to do it themselves.

Ideally a product should be maintenance free, but this is likely to involve the use of more and higher-quality components and tolerances in manufacture. This will inevitably increase the cost.

Design for Life Expectancy

Customers expect a certain minimum time that the product will work for which extends beyond the guarantee. This time will vary according to the product.

A product which fails before the end of the guarantee period is going to be very costly to the manufacturer to repair. If a product needs expensive repair soon after the guarantee expires, a customer is unlikely to make a repeat purchase of the same brand, and the brand might develop a reputation for being unreliable.

However, if a product works successfully for many years, consumers will not need to buy replacements so often, and demand will fall. The number of products made will drop, and as a result the price will rise.

Many products contain components which are likely to fail after a number of years, and which would be very expensive to repair or replace. This is known as **planned obsolescence**.

Many products have been re-designed in which mechanical components have been replaced with electronic control devices.

Spreadsheets and databases are used extensively in industry to solve many of the problems discussed on this page. A great deal of data can be recorded electronically and retrieved and manipulated when needed.

IN YOUR PROJECT

How would you set about designing your product so that it could be made:

- cheaper
- quicker
- better
- more desirable
- safer?

Designing for Manufacture (2)

The world is a dangerous place. As designers and manufacturers produce new products, they need to ensure that they will be safe to use, and also safe to make. Potential hazards must be identified and risks minimised.

There are also the potential moral and social effects of the design to consider.

Hazards and Risk

Safety First

The designer must ensure that the product conforms to all the relevant safety standards, including those of other countries in which it might be sold. Careful consideration must be given to ways in which people might misuse the product, and any necessary safety devices and warning labels included. The designer can be held responsible for any accidents which occur as a result of poor design.

There are four main areas to consider to help avoid potential accidents in the factory:

- the design of machinery and tools being used in the manufacturing process
- the physical layout of the work area
- the training of the workforce
- the safety devices and procedures.

It is also essential to reduce the number of potential hazards – unsafe acts or conditions – which could occur in the workplace. Accidents are extremely costly in terms of personal distress, compensation and lost production.

Reducing the Risks

Although we cannot avoid risk, we can take steps to assess the likelihood of something happening, and minimise its impact if it does.

As well as the legal requirements and more general codes of practice for Health and Safety, a considerable amount of documented information is available to help guide the design of safe products and working environments.

Ergonomic studies and anthropometric data can be used to determine optimum positions for displays and controls on products and machines, and the most suitable sizes and arrangements for workspaces and conditions (e.g. distribution of light, noise, heating and ventilation, etc.).

Risk Assessment

When a production process involves hazardous situations it is necessary to analyse and assess each particular risk situation and ensure that adequate precautions are taken to minimise the potential danger.

It is the responsibility of an employer to assess the risks involved in each stage of production and justify the level of precautions adopted to a Health and Safety Inspector.

Issues

When specifying the requirements for a product, designers need to consider a wide range of moral, economic, social, cultural and environmental issues. These often produce conflicts which can be hard to resolve.

Moral Issues

In certain situations a product may have the capacity to injure or harm someone – either the user or a bystander. Cigarettes and alcohol are obvious examples. Bull-bars on cars may look good and help improve sales, but they are likely to increase the severity of injury to a pedestrian in an accident.

Social issues

Some products can have a major impact on the way in which large groups of people live their lives. Convenience foods, for example, mean that there is less likelihood of the family sitting down together to eat a meal. Promotion and packaging can help counter this by providing two-person portions and using images of family meals.

Information and communication technologies are in the process of making a major impact on society, as work, entertainment and shopping can be increasingly undertaken at home. Advanced automation reduces the number of people needed to produce and distribute goods, causing unemployment.

Cultural issues

The particular beliefs and traditions of different groups of people have a major effect on the way they live their lives – what they do, where they go, and the things they buy.

Food and clothing and the symbolism of certain shapes and colours all play highly significant roles in maintaining the identity of a particular culture. When a product is intended for use by a range of cultures it is important to identify and recognise such needs.

Pager Design Specification

You will need to develop this specification according to your own investigation. Refer back to the brief on page 94 and pages 18 and 19.

Package Specification

- The target market are people such as...
- The graphic identity is to be based on...
- The device must fit a pocket, sized...
- The following controls must be included...
- The life expectancy will be...
- The maintenance requirements are...
- The number to be produced is...

IN YOUR PROJECT

When preparing a design specification statements need to be made about:

- how frequently and easily different parts of a product are likely to need to be maintained
- how long the product should remain in working order, provided it is used and maintained by the consumer as instructed
- the total number of units likely to be made, the production times, and, if appropriate, the sizes of batches.

Orthographic Drawings

Orthographic drawings combine plan and elevation drawings of an object. They provide full details of its size in three dimensions and the arrangement of the various parts.

The arrangement of the drawings and the method of dimensioning are defined by a British Standard. The standard provides a common method which everyone will be able to follow.

Arrangement of Drawings

The drawings should always be placed so that the plan and front and side elevations line up exactly. The number of views shown should be the minimum needed – usually three. This arrangement of drawings is called Third Angle Projection.

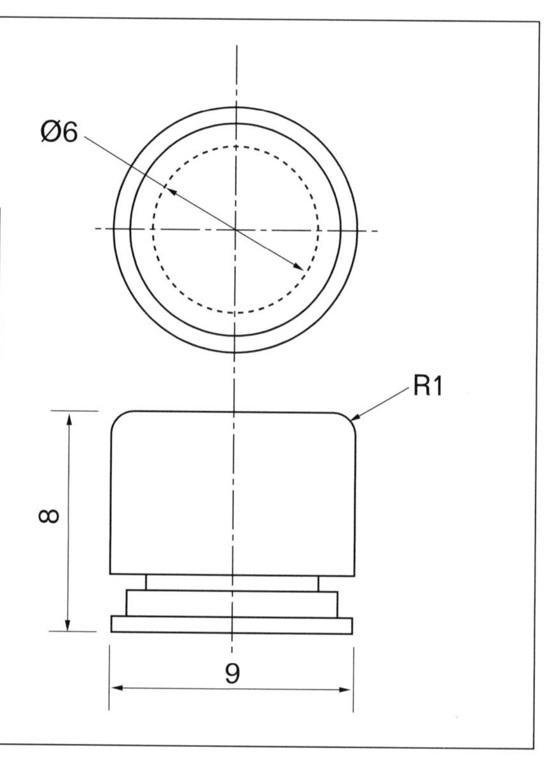

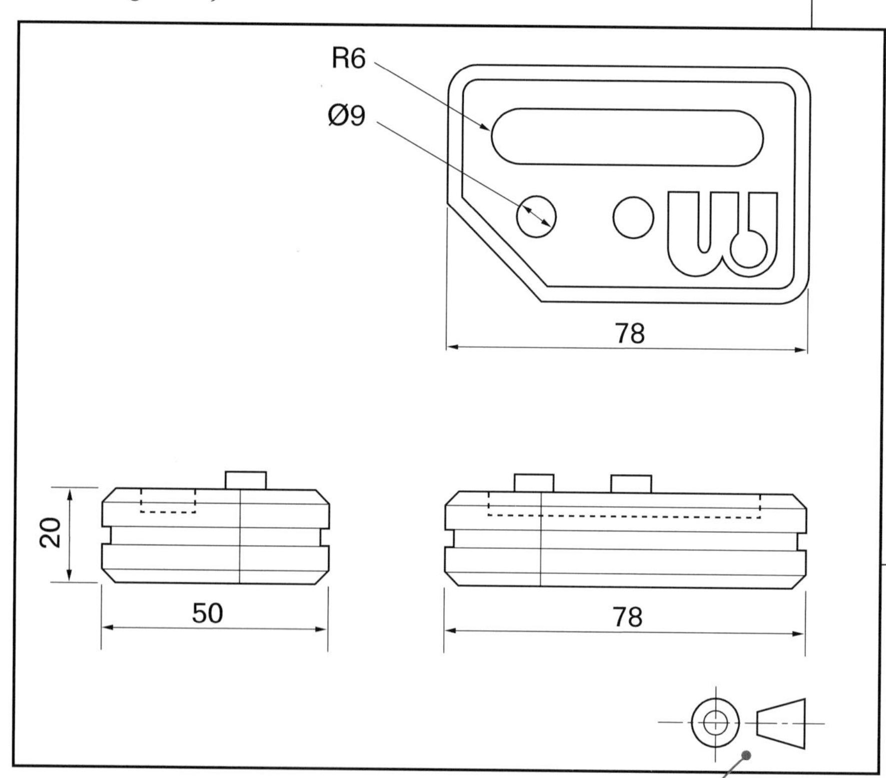

This is the British Standard symbol for Third Angle Projection.

Different Drawings

In a **general arrangement** drawing the product is drawn in its final assembled form. Overall dimensions are given, and each component is numbered.

On a series of **detail drawings**, a fully dimensioned orthographic drawing is provided for each component part.

■ ACTIVITY

Find a common object and make a detailed general assembly drawing of it. Exchange your drawing with a friend to see if an accurate copy can be made without you providing any extra verbal instructions.

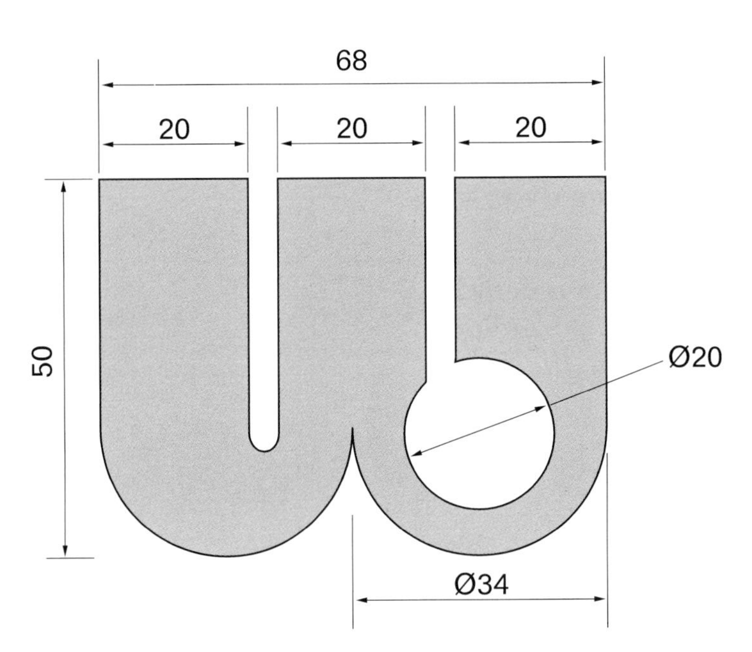

Dimensioning

There are some clear rules for adding sizes to drawings:

▷ All dimensions are given in millimetres, and are written as the number only.
▷ Numbers should be placed directly above the middle of the dimension line.
▷ Numbers should be placed so that they can be read from the bottom or from the right of the drawing only.
▷ The minimum number of dimensions should be included.
▷ Dimension and projection lines should be half the thickness of the lines of the object.
▷ There should be a small gap between the end of the projection lines and the object.
▷ Arrows should be placed accurately between the location lines, and be filled in.

Consult BS308 or PD7308 for fuller details of the system which should be followed.

Computer-aided Drawing

There is a wide range of CAD packages which enable complex and accurate orthographic drawings to be prepared. These are now extensively used in industry to save time.

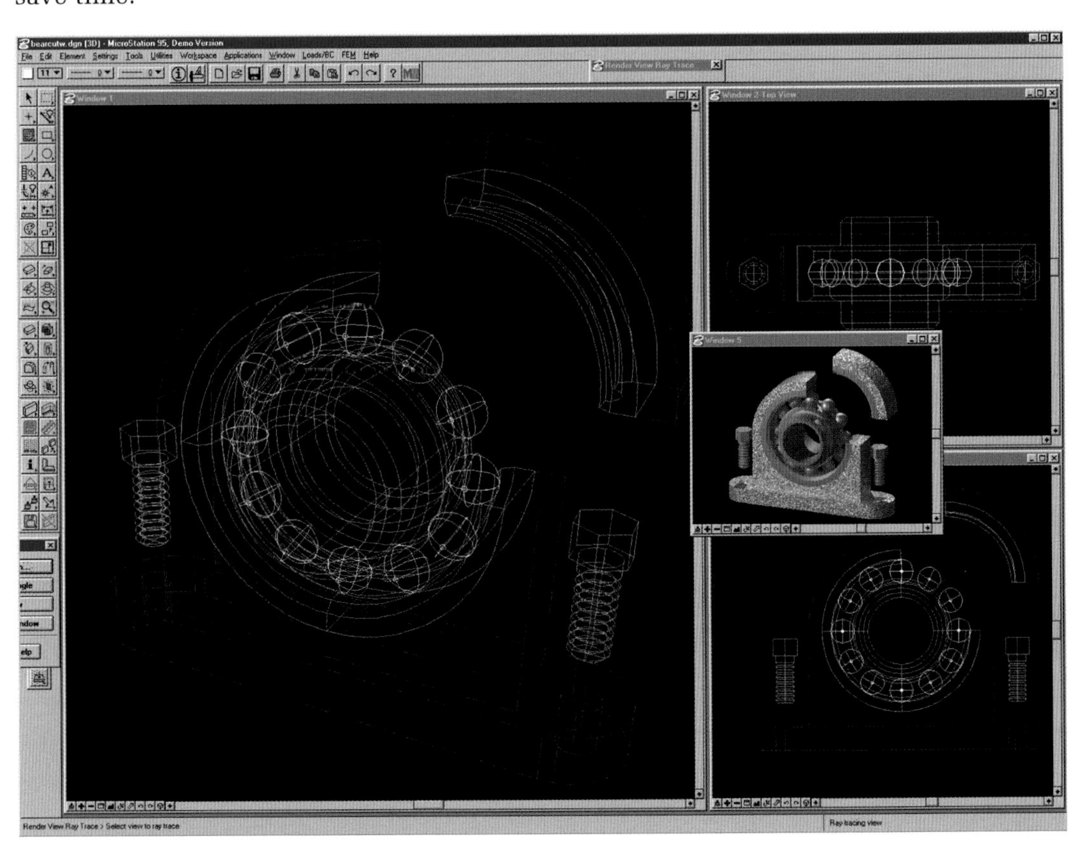

IN YOUR PROJECT

Make sure you use the approach described here when drawing and dimensioning graphic symbols or designs for products.

Try using a CAD package to prepare your drawings.

KEY POINTS

- The layout and dimensioning of orthographic drawings must follow BS308 or PD7308.

Drawing in Three Dimensions [1]

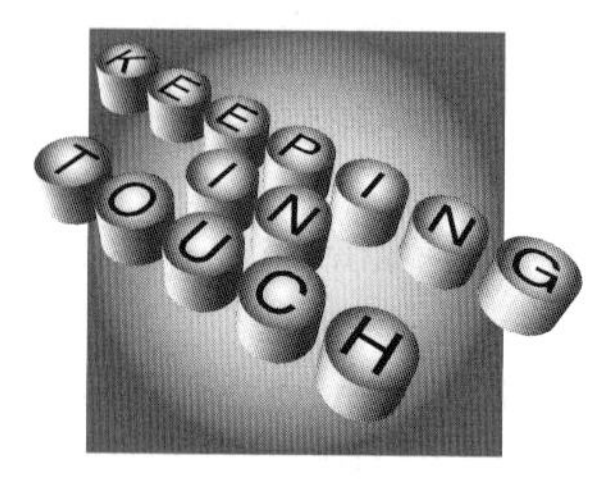

Three-dimensional drawing ranges from a tiny 'thumbnail sketch' on the back of an envelope to a carefully constructed image on a drawing board or computer screen.

Designers use such drawings to get a better idea of what the product they are designing will look like, and to communicate their proposals to a client.

Freehand Sketching

Designers need to learn to sketch their ideas quickly in freehand. This may be for their own benefit as part of the process of visual thinking and development, or it may be to show to others for discussion.

Such sketches should not be drawn with rulers of other drawing aids, as this would slow the process down, and accuracy is not needed at this stage. They are usually in pencil, ink or marker pen, but you should experiment with different graphic media to develop your own successful style.

The main thing is to sketch confidently and boldly, and not be afraid to put pen to paper, or to make mistakes! Never try to erase or cover up mistakes – just try again on another part of the page.

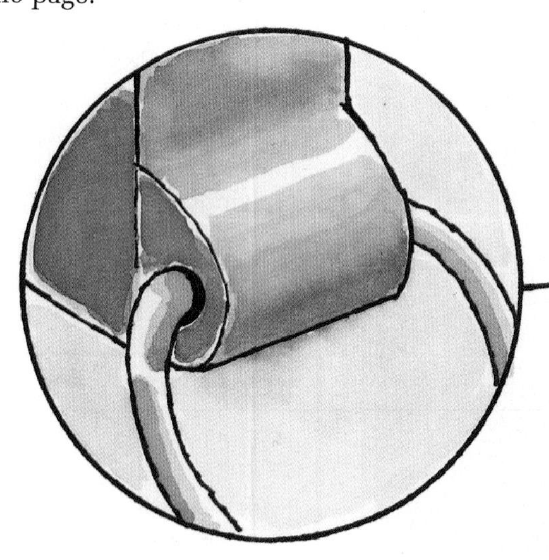

Crating

The method of 'boxing in' or **crating** is widely used to aid the construction of three-dimensional drawings.

If you can master drawing cubes, rectangular boxes and cylinders in three dimensions, it is surprising what number of everyday objects, and of course your own design ideas, you will be able to draw convincingly.

Isometric Drawing

Isometric is a drawing system that is very useful and common in product design. Although a formal isometric drawing should follow certain guidelines, it is also an excellent system for freehand sketching. Isometric drawings are fairly realistic to look at, and are quick to do. They are particularly good for use in instructional drawings.

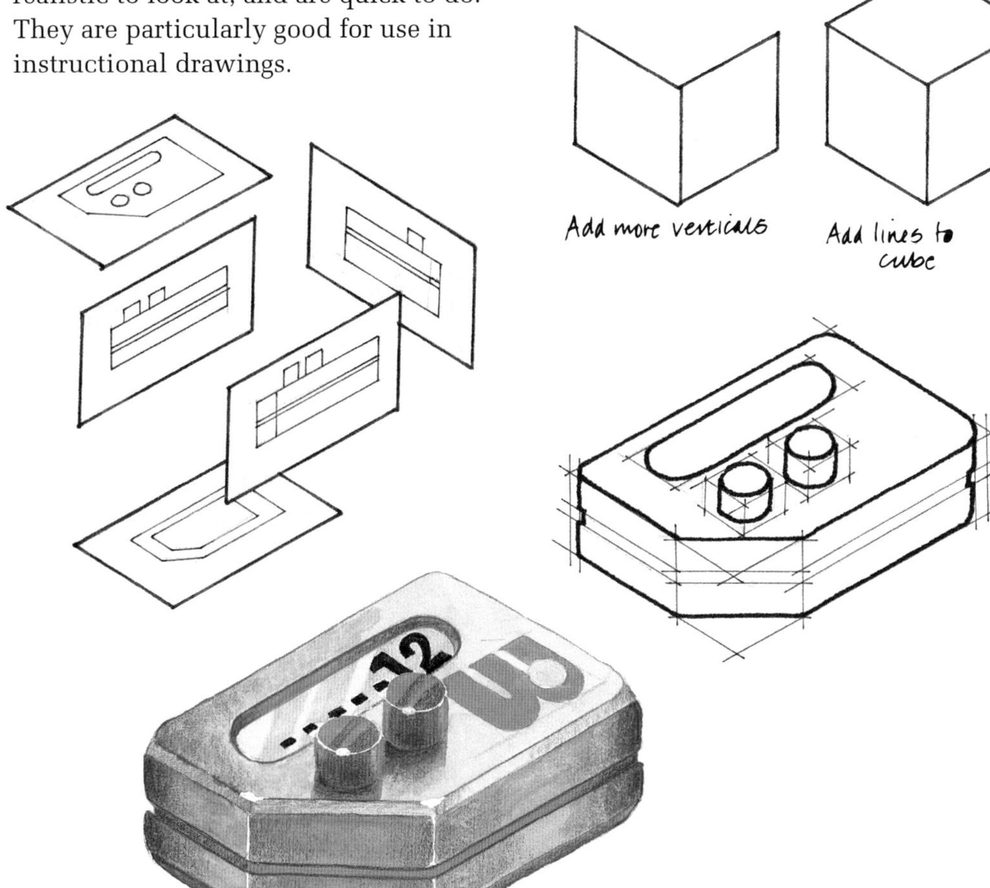

Sophisticated presentation graphics can be created using ICT – but always start by doing freehand sketches first.

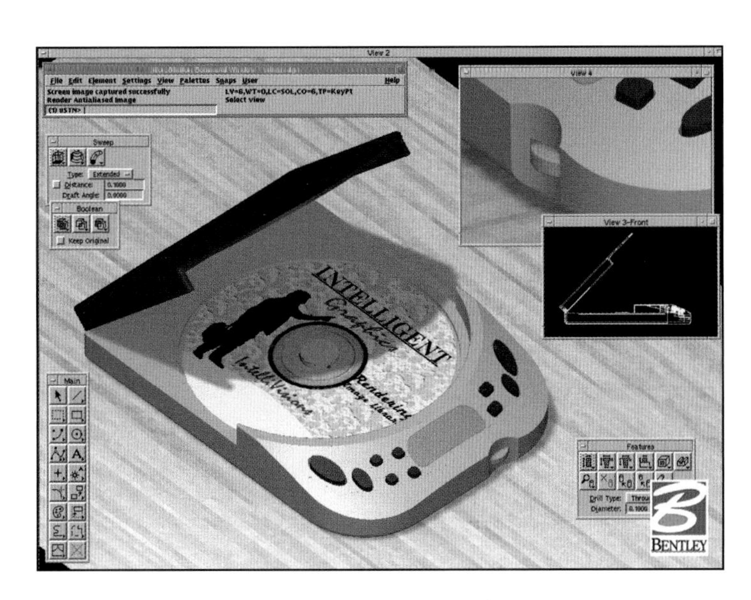

■ ACTIVITY

Obtain a familiar household object. Choose one which is mainly rectangular in shape, with perhaps some curved forms. To begin with prepare an orthographic sketch of it (see page 106).

Then attempt to develop an isometric drawing of its overall shape. Use the crating approach to develop its other shapes and forms.

Remove the constriction lines and darken the lines you need.

If you find the object too difficult to draw, choose something simpler.

If you find the object easy to draw in Isometric, move on to something more demanding.

IN YOUR PROJECT

- ▶ Use freehand isometric sketches to help develop the design of three-dimensional objects.
- ▶ Use measured isometric drawings as the basis for the final presentation of ideas for products.
- ▶ Rendering techniques (see pages 112 and 115) can be added to both sketch and formal isometric drawings

WWW.

For more information about the 3D CAD program illustrated on the left, go to: **www.bentley.com**

Drawing in Three Dimensions (2)

A cut-away drawing is a two-or three-dimensional drawing with part of the drawing left out in order to show areas that would otherwise be hidden.

Exploded drawings are used by designers to show how things fit together and how they come apart. They also show clearly details that might otherwise be hidden in a drawing.

IN YOUR PROJECT

Try to use simple cut-away and exploded drawings while sketching ideas.

Also use them as part of final presentation work to help illustrate key aspects of your design proposals.

Cut-away Drawings

When producing a cut-away drawing, the illustrator needs first to decide on the most suitable sections to leave out in order to show the detail required. Plans, elevations and isometric projections make a good starting point for isometrics.

The drawings themselves may be quite complicated to construct. Colour can be used to good effect to highlight certain details, but these types of drawing are often rendered in black line hatching to help identify different component parts.

battery

KEY POINTS

Mechanical and electrical product manufacturers often use cut-away drawings to show the workings of the things they have designed. Architects and interior designers also use them to reveal what is behind wall surfaces.

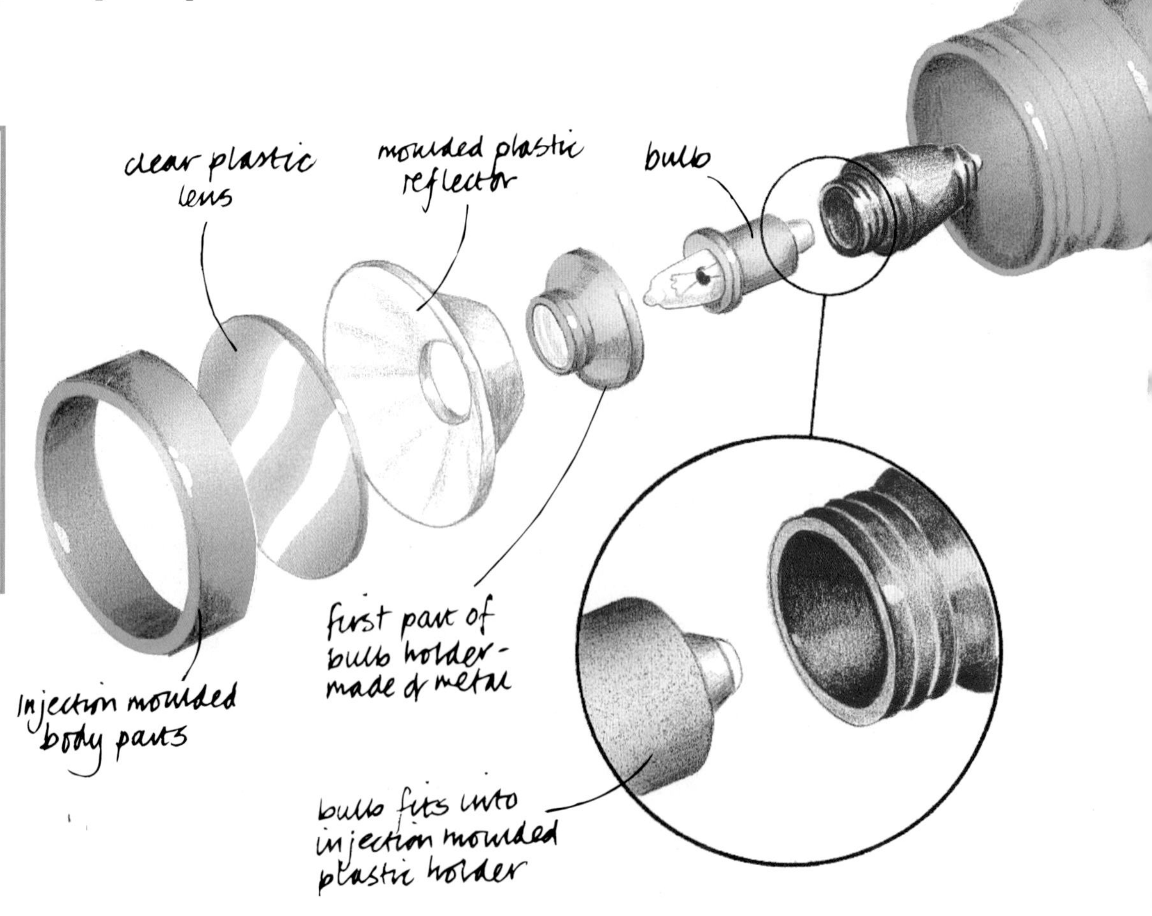

Exploded Drawings

Exploded drawings are often used in instruction or repair manuals. They help show clearly how a series of components fit and work together. An exploded drawing is usually shown in three dimensions. Isometric and axonometric projections (see page 129) are often used for this purpose.

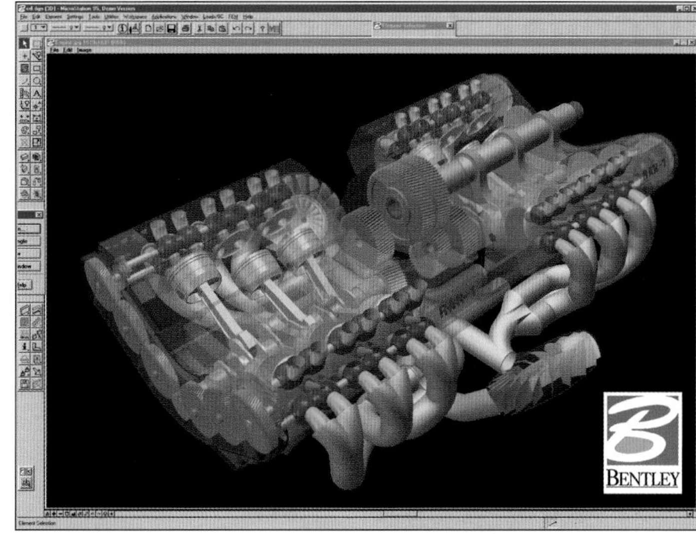

A cut-away drawing produced using CAD.

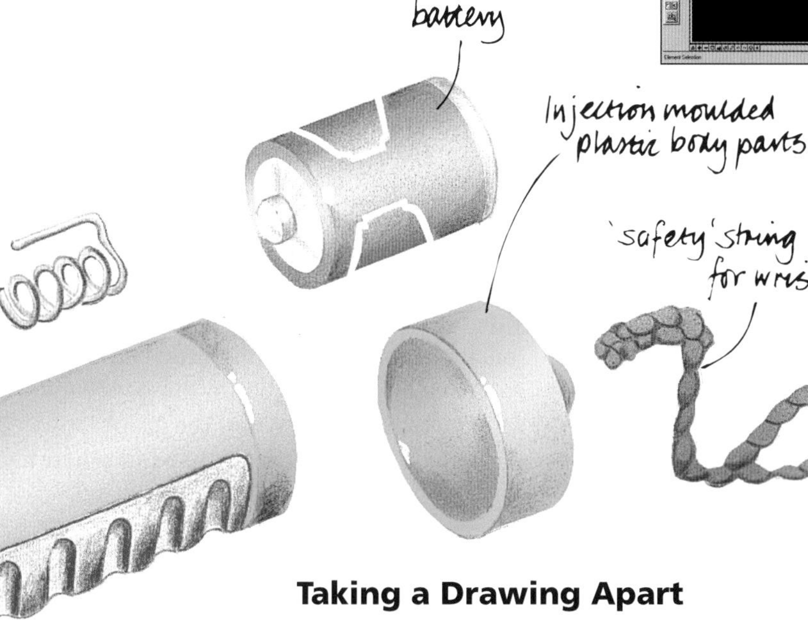

Taking a Drawing Apart

The individual parts of an exploded drawing are first drawn separately on tracing paper and moved away from each other, creating space in between the components. Careful visual positioning of the parts, using a light-box, helps makes them look as if they fit together. The separated pieces are often shown overlapping the surface they have been extracted from. The result is a dynamic and highly informative drawing.

Some important things to remember are:

- ▷ exploded drawings are not usually dimensioned, although the various parts can be numbered or annotated
- ▷ sometimes it can be effective to leave all or part of the drawing in freehand sketch form, rather than working up a highly technically controlled finish
- ▷ colour can be added in an impressionistic or diagrammatic way to help communicate different components or materials
- ▷ experiment with different textured and coloured surfaces to represent the various materials which have been used.

■ ACTIVITY

See if you can obtain an old torch. Make sure you ask permission first, and that someone has checked it is safe – torches can get rusty, and acid from old batteries may have leaked out.

Carefully disassemble as much of it as you can, by unscrewing or pulling apart the various parts of the casing.

- ▶ What does each component do, and how does it do it?
- ▶ What are the components made from?
- ▶ How do the components fit with the other parts?

Think about how the torch has been assembled, and the materials and components which have been used. Identify some of its features which you think have been well designed and some which you feel are not very successful.

Present your findings by means of an annotated A3 presentation sheet.

Product Rendering (1)

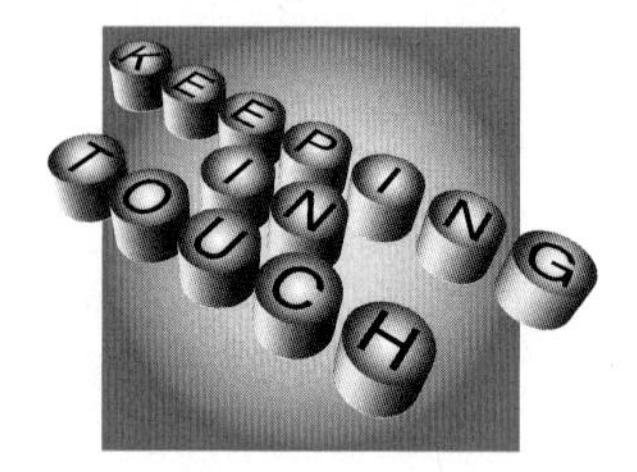

Product rendering is the process undertaken to provide a suitable finish for the presentation of a design proposal. A rough 'sketchy' style, or something more accurate and highly finished might be required.

To find out more about the history of graphic media, go to **www.berol.co.uk**

Pencils

Graphite pencils are available in different hardnesses of lead ranging from 9H at the hard end of the scale to 9B at the soft.

Hard leads are excellent for technical work as they are precise and stay sharp for longer. They are therefore good for architectural-type drawings and orthographic drawings of products. They are also accurate for tracing. One drawback is that they leave an impression on the paper that remains even when the graphite has been erased. So care is needed when using a hard lead, and a light touch.

One tip is to draw guide-lines or marks with a softer pencil and only use the hard pencil when the final drawing has been worked out.

The softer pencils are ideal for sketching both from observation and initial ideas. They lend themselves to a freer approach and are excellent for applying shading, tone and shadows to a drawing. The side of the lead is particularly good for this. The really soft pencils will need a lot of sharpening!

Coloured Pencils

Coloured pencils are available in a large range of colours and are easy to control. They are good for textural results as well as 'flat' colouring, although the latter takes practice to achieve.

Coloured pencils are not normally available in a variety of soft or hard leads as graphite pencils are although the better quality ones tend to have softer leads.

An excellent use of coloured pencils is to overlay colours and gradually 'build' the colour you want. They are used effectively in mixed-media renderings for highlighting, detailing and shadows. Water-soluble colour pencils are available which can be 'blended' and 'graduated' with the addition of water.

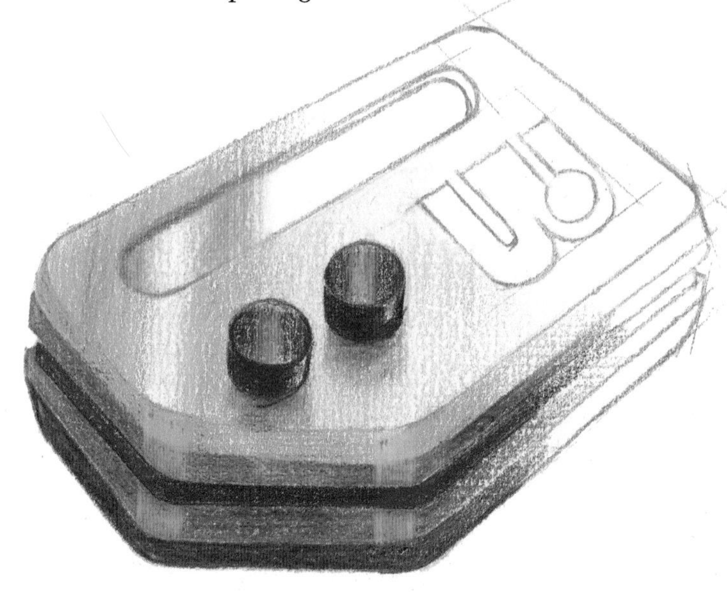

Felt Markers

Coloured felt markers are widely available and probably the most common media used by the graphic designer. They are sold in an enormous range of sizes, colours, styles and prices. Some have become positively fashionable!

Cheaper pens tend to be water based, and when used carefully can produce good results. It is difficult to render flat areas of colour with these pens because they dry quickly and 'streak' where the lines of colour overlap. Use them boldly and cover the area only once for the best results. Constant overlaying leads to the paper becoming saturated and eventually breaking up. Water-based pens are excellent for smaller areas of colour and for lettering and fine detail.

The best markers are without doubt spirit-based ones. These are widely available in a huge range of colours and are the pens used by professional designers. They tend to be expensive, so one tip is to supplement your usual range of pens with a few of these markers – perhaps some primary colours and certainly a range of greys.

These pens require practice to achieve the slick results often seen in books and certainly there is a knack in using them. Spirit-based pens tend to 'bleed' on most papers and allowance should be made for this by not colouring right to the edge of the image but leaving the bleed to do this for you. Practice will make perfect!

Technical Pens

Technical pens are used for drawing fine, exact lines. They are best suited to architectural plans and accurate design drawings such as orthographic layouts. The best quality pens are expensive and can be taken apart for cleaning. They have a wide range of accessories available such as compasses, stencils and nibs.

Special inks have been developed that are permanent and densely black. Colour inks are also available.

Cheaper, disposable technical pens are good value, but do not last as long.

A feature of technical pens is their square-cut hollow nibs. These are not that suitable for holding at an angle to the paper and therefore not the best choice for spontaneous freehand sketching.

The Airbrush

Originally invented for retouching photographs, the airbrush uses a low-pressure air supply to blow a fine spray of ink onto an image area. It has become a sophisticated tool. Airbrushed, photographic-quality images are familiar on record sleeves, posters and exploded drawings.

Although the airbrush is too time consuming for rough visuals, they can be used for high quality illustration work. It requires a lot of patience and practice to create good results with an airbrush. A lot of masking is needed to prevent the spray getting on to the wrong areas. Flat areas of colour can be produced, but the airbrush is most effective in rendering graduated tones.

Product Rendering (2)

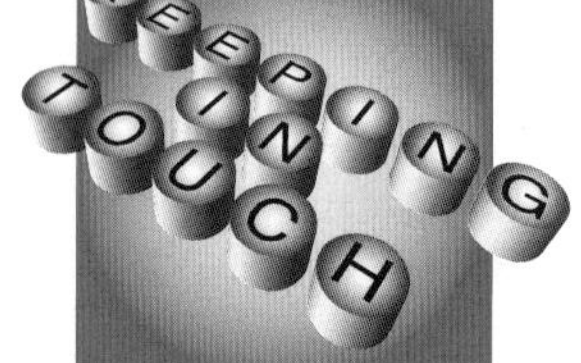

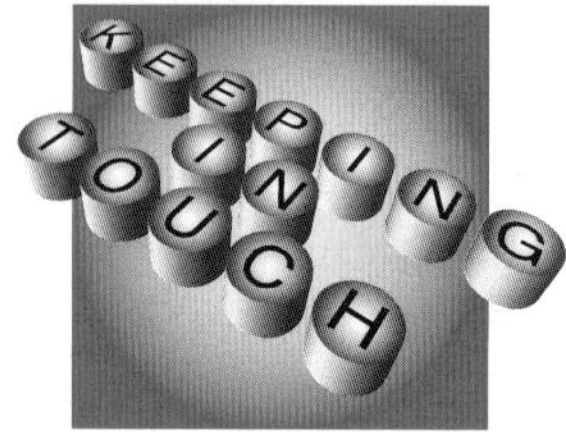

Water Colour

Water colours are best used as a 'wash', leaving the paper to show through to create the lighter areas of colour on the drawing. They are good for subtle colouring, perhaps combined with strong, detailed linework. They can be mixed and blended on the paper giving good graduations of colour. Water colours are widely available in tubes or pallets.

Gouache

Designers' gouache is a more expensive paint, used in a thick, creamy consistency to achieve opaque colours. The colours tend to dry slightly lighter than when mixed. It is possible to paint areas of opaque white and therefore is particularly good for highlights. Gouache is easy to control and is particularly good for producing areas of flat colour with accurate edges.

Acrylics

Acrylics are plastic-based paints which provide bright opaque and wash colours. They are used by artists as a painting medium on paper, board and canvas but also by designers for painting three-dimensional designs, such as models. They have the advantage of being good for use on most materials, including plastic.

Spray Paints

Spray paints, such as those used for car repairs, are excellent for colouring models. It is important to be careful spraying on to some materials such as polystyrene, and it is worth experimenting first. Be sure to follow the safety instructions provided.

Enamels

These are good for painting models and provide a hard, glossy, knock-proof surface. They are available in a wide range of colours from model shops.

Chalks and Pastels

For the graphic designer, these are ideal choices for adding tone, shading and broad areas of colour to visuals. They can be effectively blended and graduated and used in combination with pencils and markers as they are not particularly good for rendering fine detail. They are one of the few graphic media that are successful when used on coloured paper or board backgrounds.

■ ACTIVITY

Experiment to find a good way of getting a really crisp edge to your rubbing out.

Erasure shields can be used to cover areas that you want to protect. Any piece of paper or card can be used, although you can buy plastic ones.

Light and Dark Tones

Shading can simply be a pattern of dots or dashes or lines which can be cross-hatched to provide a variety of intensities. More continuous tones can be achieved with the side of a pencil or with paint. Rub-down coloured film and textured sheets can be very effective; the latter are used a lot in newspaper graphics because its reproduction qualities are particularly good and they are quick to prepare.

Highlights and Shadows

Highlights are normally very effective on graphic renderings. They provide a touch of realism, particularly on three-dimensional drawings. A white pencil or a blob of white paint at a critical point works wonders!

On certain art boards the rendered colour can be scratched away or erased to reveal the white surface underneath. On pencil drawings a good quality eraser can be used to create highlights. It is fairly difficult to remove most colours once applied so you will end up putting the highlights over the top. Experiment to find which works best.

Simple shadows are easy to create, but accurate shadowing needs to be worked out using technical drawing methods and can be complicated, and is certainly time consuming. A good method is to fix an imaginary light source and project your shadows from that. Set up objects under an anglepoise lamp and observe shadows. See where they are at their most intense and see how they fade as they get further from the light source. Shadows do need to be consistent to be effective.

ACTIVITY

Obtain a range of rendering materials and equipment that you are unfamiliar with.

Draw a cube, a cylinder and a sphere and experiment freely to discover what works well, and to begin to develop your own personal style.

Add tones, highlights and shadows to your renderings. Try to represent different textures and surfaces of a range of materials, such as plastic, wood, metal and glass.

IN YOUR PROJECT

- How sketchy or technically accurate does your rendering need to be?
- What drawing instruments and media would be most effective?

KEY POINTS

Rendering can be used for a number of reasons:

- to provide an impression of the three-dimensional form of a product
- to indicate materials and textures of surface finishes
- to show colour.

Product Modelling

Designers use 3D presentation models to help communicate their ideas to their clients. Sometimes the models work, but their main purpose is to show what a product or space will look and feel like in real life. Model making is a specialised skill in its own right.

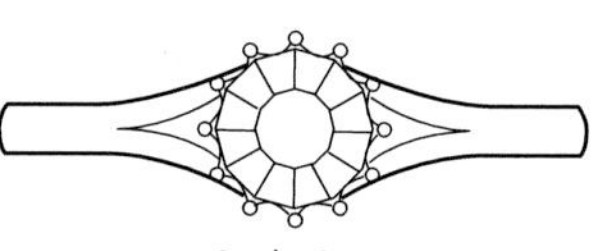

Scale 2:1

Scale 1:1

Scale 1:500

Scale

One of the first decisions is what size to make the model. Here are some commonly used scales:

- ▷ A small intricate item – 2:1 (i.e. twice as big)
- ▷ A hand-held object – 1:1 (i.e. full size)
- ▷ A piece of furniture – 1:10
- ▷ An exhibition stand or a room – 1:50
- ▷ A house and garden – 1:100
- ▷ A public building – 1:500.

At a scale of 1:10, 1 unit represents 10 units in reality. A model which is 100 mm tall would show an object which is actually 1000 mm high.

It is a good idea to check the sizes and thicknesses of materials you have available. These might influence the exact scale you end up choosing.

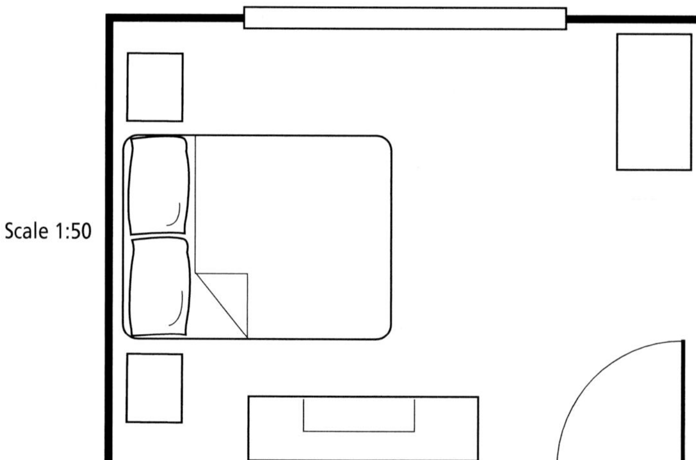

Scale 1:50

Materials

Most presentation models use completely different materials to those which would be used for the final product. You may need to adapt old bits and pieces of discarded objects by reshaping and repainting them. Some of the best model-making materials are:

- ▷ dense styrofoam
- ▷ dowel rod
- ▷ drinking straws
- ▷ wire
- ▷ papier mache
- ▷ card
- ▷ vacuum-forming plastic

Finishing Touches

It's often the attention to details which really brings a model alive. Carefully modelled displays and controls, textured surfaces and applied graphics all help make it look and feel real. If the actual product would be moulded, add grooves to show where the different parts would be fitted together.

If you can, paint the model using spray paints (make sure you follow safety requirements). You will need to rub the surface down and respray it several times to get a really smooth finish.

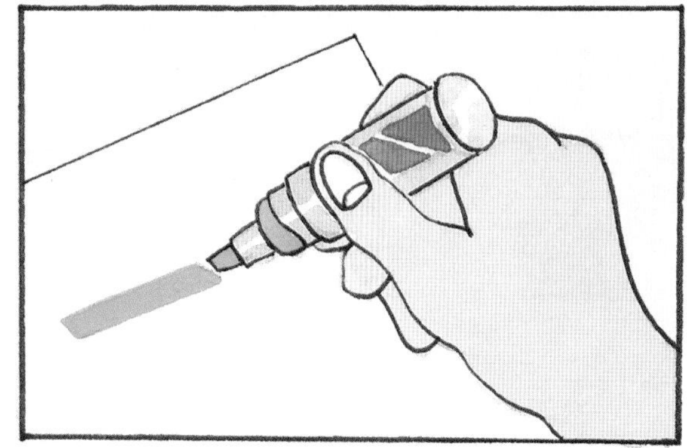

ACTIVITY

What scale would you make a model of each of the following:

- a postage stamp
- a portable telephone
- a dining table
- a stage set
- a shopping centre?

What model-making materials might be particularly useful for each?

IN YOUR PROJECT

Think carefully about what key aspects of your design ideas you need to present through your model.

What will best be communicated through:

- a 3D model
- a series of drawings
- words and numbers?

Setting the Scene

If you have made a full-size model, photographing the product being used by someone in the setting it would be used in will make it look even more convincing.

If you have made a scale model, draw, or find a photograph of, a person or a familiar object which is at the same scale.

Adding a background to a model of a building or outdoor structure can sometimes be very effective.

KEY POINTS

When planning a model, designers need to think about:

- choosing the right scale and selecting the best materials
- achieving a high-quality finish
- showing the model in use.

3D Computer-aided Design

CAD systems make it easy to create a picture of the intended product, and then to change how it looks on screen. This means that new ideas can be tried out and evaluated much more quickly than if they had to be redrawn each time.

There are a number of different types of CAD program available. Each is best suited for modelling different types of products and materials. Some are more concerned with visual appearance, others with technical detail.

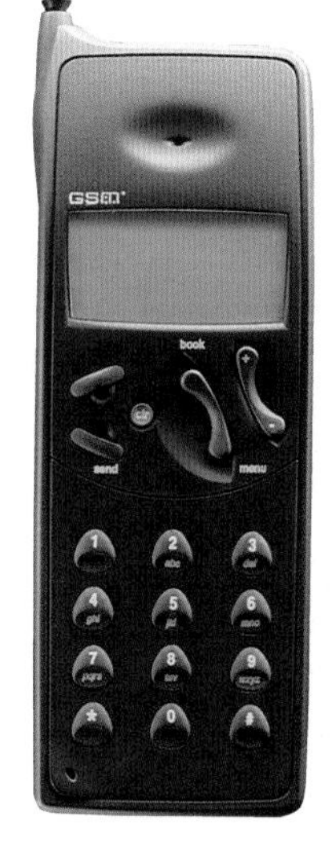

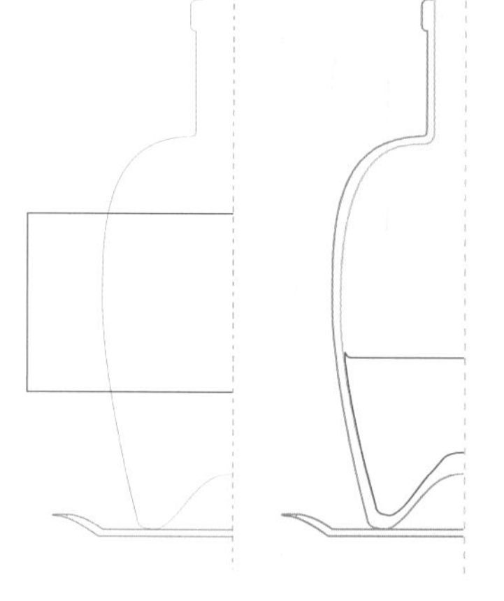

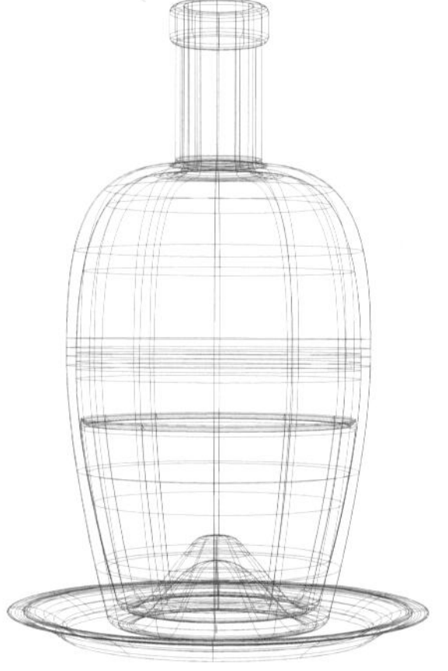

Methods of Representation

There are three main ways of representing an object on a computer.

▷ **Wire-frame modelling** means that the object is represented by a series of lines. This image can be enhanced by removing lines which would be hidden.

▷ **Surface modelling** can then be added. The surfaces of the object are represented by colour, shading and texture to give a stronger sense of the 3D form.

▷ **Solid modelling** means that the drawing is based on geometric shapes which can be mathematically analysed. These can provide information on such things as the object's mass, volume and centre of gravity.

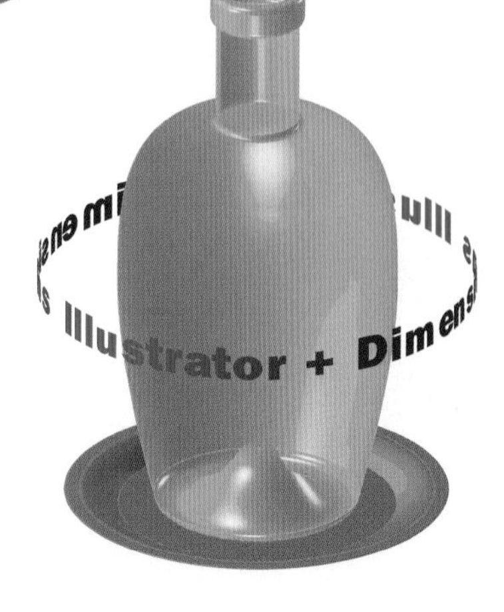

Transforming a CAD Model

Most CAD programs have certain common facilities. Drawing tools produce lines which can be adjusted for width, shape and colour, and can be set to produce curved lines, or geometric shapes.

Text can be added in a range of styles and sizes. Measuring systems provide accurate information on lengths of lines and their position on the page.

It is very easy to transform a CAD drawing or image by copying, rotating and mirroring different parts of it. In some programs the drawing can also be used to analyse and predict stress points and how liquid materials would behave inside its form.

AKA Mobile Phone

The first stage was to create two appearance proposals for the design of this new portable telephone. These were presented as hand-made models. The models were made from a material called Ureol which is a resin that can be cut and finished easily by hand.

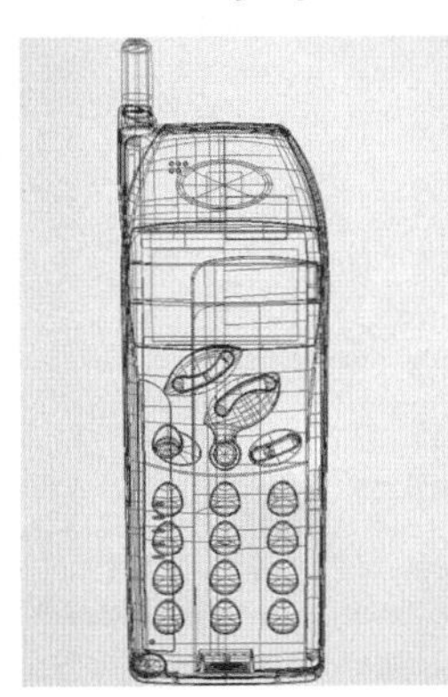

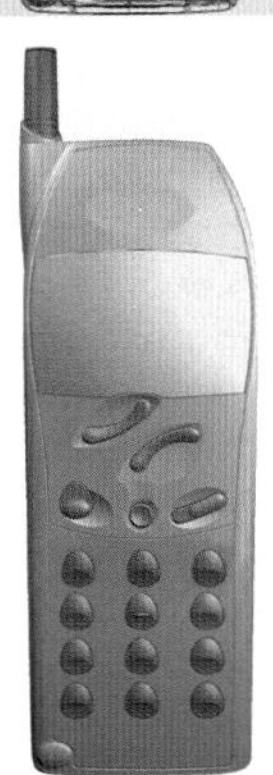

Next AKA were commissioned to combine the designs and the technical specification into a viable product. Alias software was used to create digital models of all the electronic components and to then wrap a 3D 'skin' round the. Buttons were positioned to ensure existing components could be used, saving development time and costs.

Shaded images of the wireframe were constantly assessed to ensure that the model met the aesthetic as well as technical requirements.

AKA then created a physical model to present to the client. This one was made directly from the computer data using a CNC milling machine.

A Rapid Prototype was produced to check that all the electronic and components and plastic housing fitted correctly. This was created using a laser cutting a liquid resin, again fully computer controlled from the original computer data.

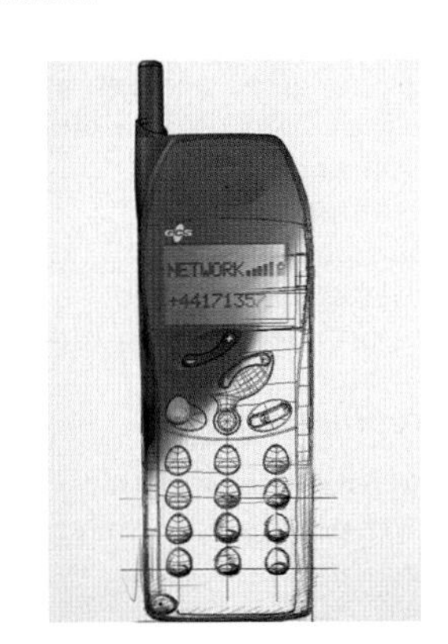

Fully rendered colour images were then created. Colours, textures and graphics were assigned to the surfaces of the digital model and photorealistic images created. These pictures were used in broachures and advertsing posters for the product launch.

Finally the 3D digital data was sent via modem directly to a toolmaker in Seoul, Korea where the production tools were made.

Rapid Prototyping

Rapid prototyping is a term used to describe the potential of being able to construct a real three-dimensional appearance model quickly and easily from the computer model data.

Rather than progressively removing materials from a solid block until the shape emerges, the rapid prototype is built up in a series of highly accurate cross-sectional layers, avoiding the need to make complex tools. All these innovations radically cut development times and costs, which is essential to survival in the market place.

Virtual Prototyping

A **virtual prototype** is a very high-quality computer-generated image of a product, viewed on screen or printed out, which provides a photo-realistic impression of what it will look like when made. These are used to help communicate a design idea to the client in a convincing way. The great advantage over the traditional airbrush is that it is easy to make detail changes without the need to recreate the entire image.

Making It Somewhere Else

New ways of creating and sending electronic information are being developed. This means that the final design data can be sent anywhere in the world almost instantaneously. It can then be fed directly from the computer into manufacturing equipment to make the product.

ACTIVITY

Find out what CAD hardware and software is available in your school, and what it can and can not do.

- What advantages would there be in using CAD?
- What disadvantages might there be?

KEY POINTS

The main advantages of using CAD are:

- speeding up the process of design development, and as a result the number of designs which can be produced
- an improvement in quality of design, because the computer can more accurately simulate and produce information about how a design will behave in different operating conditions
- any changes made can be quickly communicated throughout the team working on the design
- design information generated on the system can be stored on electronic media (e.g. hard discs, floppy disc CD ROM) and quickly and easily retrieved at a later date.

Making It by Computer

More and more production processes can now be done by machines which are controlled by computers. Automated manufacturing is safer, quicker, more reliable and, in the long run, cheaper.

Computer-aided Manufacture

Computer-aided manufacture (CAM) is a term used to describe the process whereby parts of products are manufactured by equipment which is controlled by a computer.

One of the restrictions of batch production is that after a relatively small number of products have been made a machine has to be reset to the requirements of a different product.

The main advantage of CAM is that the new instructions are stored electronically and can be down-loaded and programmed into the machine very quickly. This also facilitates making small changes to the design to suit changes in the market or to produce specialised short-run products for individual clients.

Where computer-aided manufacture is used to replace a manual operation, greater productivity is possible, because the machine can work continuously. There is also a greater consistency of quality, and fewer faulty goods produced. CAM systems can also work with materials and chemicals which might be harmful to human operators.

Computer Numerical Control

Computer numerically controlled (CNC) machine tools can be independently programmed, but also have the facility to transfer data with other computers. They therefore become part of a complex automated production system.

Industrial robots can be programmed to perform a variety of tasks using a range of tools and materials. Automated and autonomously guided vehicles can be used to transport components, tools and materials to the appropriate machine.

These robots look nothing like the people they replace.

Computer Integrated Manufacture

CAD and CAM systems contribute to the development of computer integrated manufacture (CIM). This is the concept of a totally automated production process with every aspect of manufacture controlled by computer.

Powerful CAD systems can be linked directly into CIM systems. This means that the entire design development, production scheduling and manufacturing operation is undertaken by a single system. Manufacturing companies which have adopted such systems have been able to make dramatic reductions in the time it takes them to design and make their products. They have also been able to increase their quality and reliability.

Components for Rolls-Royce aero-engines are precision-machined and assembled using a state-of-the-art Advanced Integrated Manufacturing System.

Presentation Matters (1)

Good presentation of design ideas is essential. Display panels need to be clearly set out and visually interesting to attract attention and encourage the reader to examine them more closely.

There are various simple techniques and tools which can be used to help achieve this.

Planning Your Presentation

Experiment by trying out some of the following suggestions:

- ▷ Break up a plain white surface by using strips of coloured and textured paper.
- ▷ Use exaggerated torn-paper effects.
- ▷ Position titles and explanatory diagrams on unusually-shaped cut-outs.
- ▷ Develop a co-ordinated colour scheme which works across all your panels.
- ▷ Print (using a computer or photocopier) or write on tracing paper and place over images and coloured papers placed underneath.
- ▷ Blow-up details of images on a photocopier.
- ▷ Montage/overlap photos together.
- ▷ Carefully cut out images completely from their backgrounds.
- ▷ Devise a numbering system, logo, and perhaps a border.

Presentation Matters

IN YOUR PROJECT

You will find the suggestions for presentation particularly helpful when presenting investigation work and when preparing final realisation panels to communicate your design proposals.

- ► Start by deciding on the main subject of each panel, and state this clearly with a main title.
- ► Work out the key ideas you want to get across on each sheet, and make sure these come over clearly.

Presentation Matters (2)

The Photocopier

The photocopier is an invaluable tool for the designer. It enables design ideas to be explored and presented quickly and easily.

Most copiers reduce and enlarge and have adjustments for tone. They are ideal for altering the size of images for tracing purposes, and collecting a range of images for use as reference material.

Excellent collages can be made from photocopied images both for research material and as artwork in its own right. Try colouring over the top of photocopied sheets with pencils or markers for interesting effects. Be careful with spirit-based markers as they may remove some of the ink from the photocopy and leave it on the end of the marker!

Photocopying onto clear acetate film makes good overlays, particularly for type. They are excellent time-saving devices for reproducing blocks of text for mock-up leaflets, magazines and items like bar codes and ingredients for packaging.

Although more expensive, coloured photocopies are now widely available and the results are excellent. They are well worth trying for the final piece of presentation artwork.

■ ACTIVITY

Try:

- ▶ cropping, over-lapping, masking and re-photocopying images
- ▶ mixing photographs and drawings and then re-copying
- ▶ using tracing paper to 'soften' an image
- ▶ moving the image slightly while being copied
- ▶ copying onto coloured paper.

ICT

A computer 'Presentation' program such as *PowerPoint* can be very effective. Think about how you can combine animated text, graphics, photographs and sound. The best presentations are usually the simplest.

Photography

A digital or 35mm film camera is particularly useful for a designer, for example for:

- ▷ gathering research material on location for use at a later date in the studio
- ▷ recording the various developmental stages of a design project
- ▷ photographing final presentation models and artwork.

Three-dimensional products or models can be set up under studio lights and photographed for presentation purposes. Remember to provide some indication of scale.

Although most photographs tend to be in colour, don't ignore the atmospheric possibilities of black and white film. You may have access to a photographic darkroom or computer photo-manipulation program which will give a satisfying amount of control over your final prints.

Polaroid and other 'instant' cameras have the advantage of speed in that they save the processing time for prints. This enables the designer to work from the results immediately, but the quality of the image obtained is not as good as from a 35 mm camera, however, and the cost of the film is much higher.

Planning and Making It

Planning the Presentation

How are you going to convince In Touch Telecommunications that your design ideas are good enough for them to invest millions in setting up production, promotion and distribution?

What methods of communication can you use most effectively?

On page 95 you were given a list of information and ideas that In Touch Telecommunications had asked you to present to them. Look back at the specific requirements now and decide what materials you will need to prepare:

- ▷ a range of self-explanatory display panels
- ▷ a product model
- ▷ a technical report.

Check exactly how much time you have to complete this work.

- ▷ How will you divide your time up?
- ▷ What are the priorities?

Prepare a plan of action. Keep a diary record of how things go, noting the changes you needed to make as you went along.

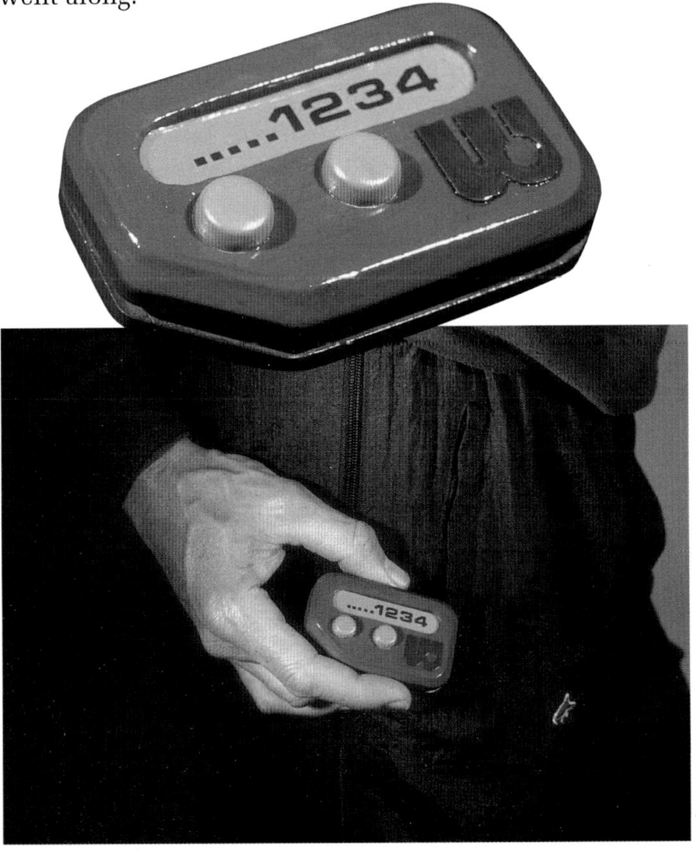

Final Testing

Show your work to some people who have not seen your ideas before.

Identify some key questions to ask them to discover how well they understand your ideas. For example:

- ▷ What sort of people is the pager aimed at?
- ▷ What words would you use to describe its shape, colours and graphic details?
- ▷ Can you tell what the controls and displays do?
- ▷ What are its key design features?

As these questions are asked, try to notice which display panels the viewer is looking at to find the answers.

- ▷ Which drawings provide the most helpful information?
- ▷ What information is missing?

Also invite open comments and questions.

Final Evaluation

In your final evaluation remember to write about both the *process* you went through and the *product* you created (i.e. the final client presentation panels). Refer directly to your final testing when discussing the successes and failures of your product.

Project Four: Introduction

Using Your Imagination (page 136)

It's not much use designing and manufacturing a great idea if no one knows it exists. An important part of a promotional campaign is the product launch, and the leaflet or brochure which people will take away. A promotional film or video can also be a very effective way of telling potential customers what the benefits of a new product or service are.

Designing Leaflets and Brochures (page 134)

The Product

Acme Accessories manufacture a range of sports-related accessories. Their latest products have been visually co-ordinated and will be sold under the brand name of 'Winner'.

You have been asked to devise a promotional campaign to launch the new product range. Your proposals need to include designs for the layout and appearance of a transportable trade/public exhibition to fit a given space and to meet other specific requirements. You are also asked to propose ideas for a leaflet or brochure, and a promotional video or computer-graphic presentation to be shown on the stand.

The Presentation

To do this you will need to prepare the following materials for a presentation session to Acme Accessories:

▷ a floor-plan and 3D drawing of an exhibition stand, including an indication of the colours and textures of the materials used for display surfaces, flooring and other fittings

and one or more of the following:

▷ the final artwork for a leaflet or brochure about the product
▷ an advertisement for the product to be used on the display stand, and in colour supplements
▷ a promotional storyboard and video
▷ a computer-animated presentation.

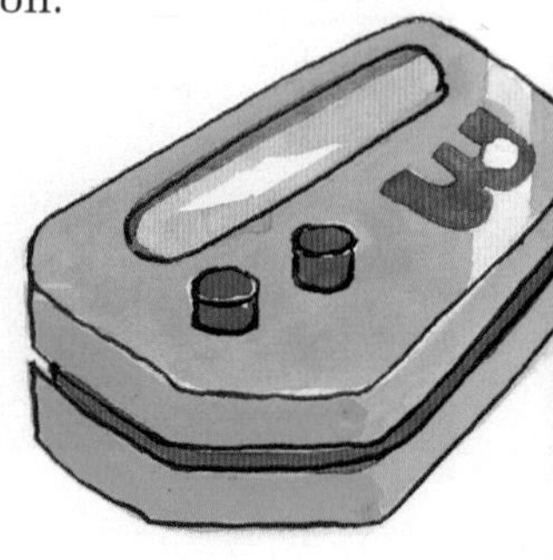

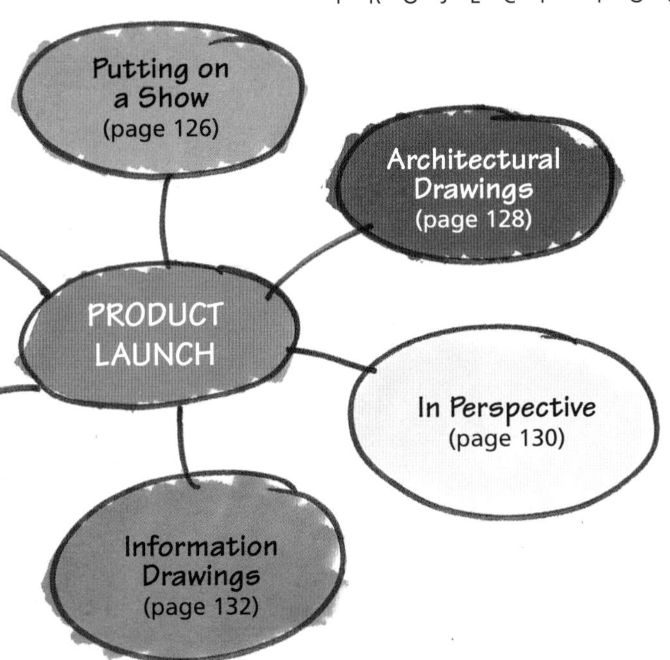

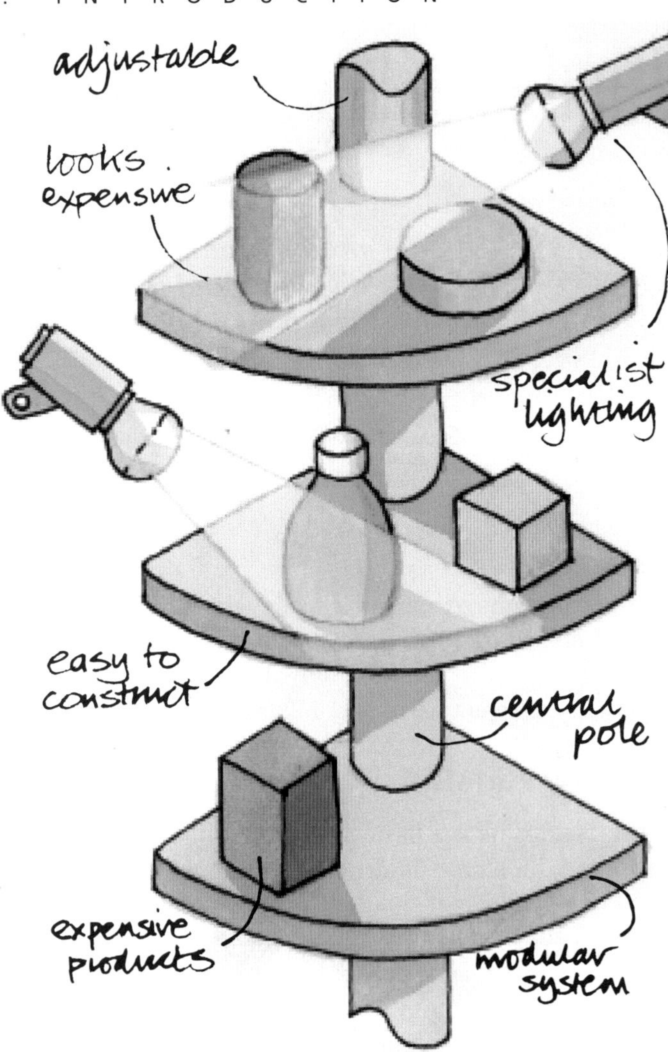

On Display

Visit either a local museum or a large trade show, or choose an interesting shop window – somewhere where there is an interesting display of 2D and 3D items.

Look closely at, and sketch part of, a display. Look out for the materials and methods of construction used for display stands. Make notes of colours, textures and special lighting.

Spend some time watching people looking at the displays.

- ▷ Which ones seem to attract their attention first?
- ▷ Which other ones do they look at?

How long do they spend looking at each display?

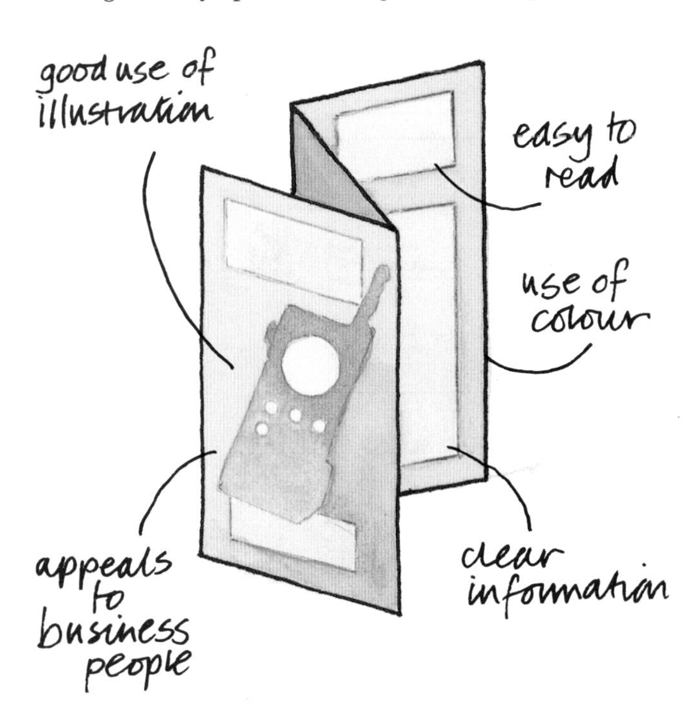

Please Take One...

Obtain a sales leaflet or brochure which provides you with information about one of the following:

- ▷ a domestic product (e.g. a vacuum cleaner or a camera)
- ▷ a mobile telephone
- ▷ a local hotel
- ▷ a local tourist attraction
- ▷ some other product or service.

Refer to the design requirements on pages 134 and 135. Make notes about the way the text and illustrations are laid out on the format. See also pages 134–135.

- ▷ How well do you feel it communicates information about the product or service?
- ▷ Is anything missing or not clear?
- ▷ Why have the illustrations used been included?
- ▷ What does it say about the product and service and the sort of people who might want to use it?
- ▷ Does the design help persuade you to want to pick it up and look through it?
- ▷ Does it tell you how and where you can obtain further information?

Putting On a Show

As part of the launch of their new Winner product range, Acme Accessories need a display stand for a large trade show which members of the public are also admitted to.

The Space

Acme Accessories have booked a space which is 6 by 4 metres. It is closed on the left- and right-hand sides and at the back, but open at the front and at the top. The tallest part of the structure may not exceed 4 metres.

The Display Panels

Acme Accessories have a basic system of 10 light-weight display units which are 1 metre wide by 2.5 metres high. They also have a range of smaller boards, display tables and cabinets, etc. The boards can be placed at any angle, and can be used on one or both sides.

3D Displays

There will need to be at least one display show-case to present the products, and a means of containing take-away leaflets.

TV Presentation

There will also need to be at least one large TV screen (1 metre by 1.3 metres) on which a promotional video can be shown. The screen can be built in to one of the display units at a suitable height.

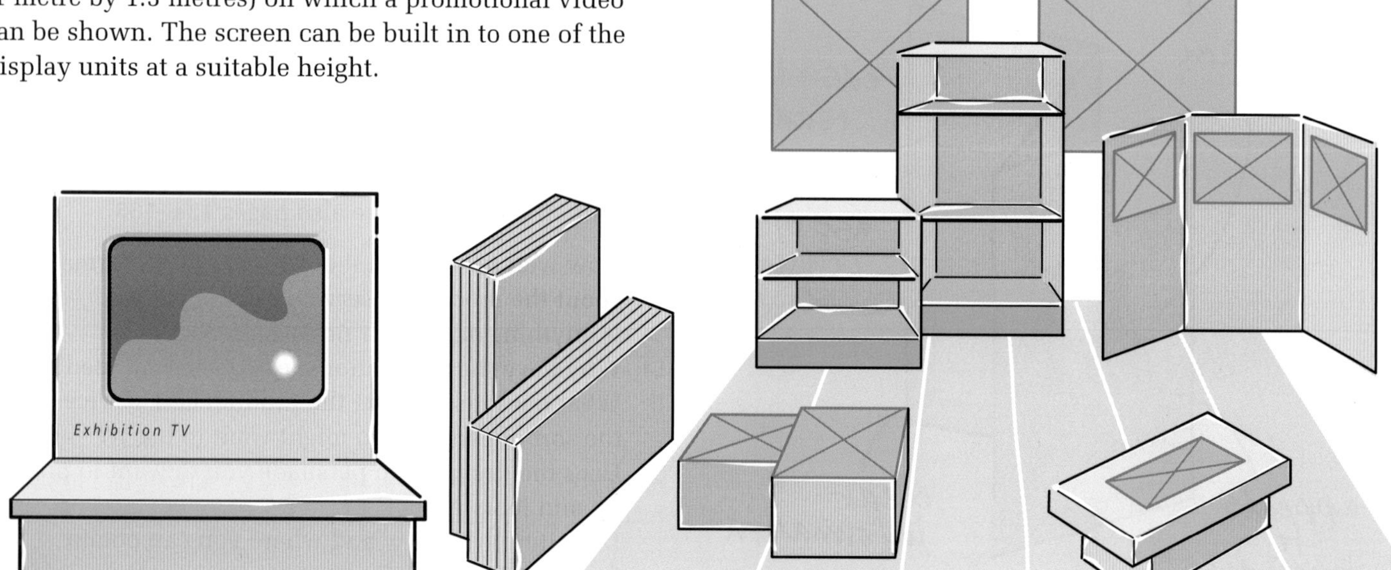

First Thoughts

Sketch out some plan arrangements of the display boards in the available space. How can you make the stand interesting and distinctive? Try to find a way of drawing passers-by in.

Don't forget to think of the needs of the people who will supervise the stand.

Developing Ideas

Develop some of your more successful ideas by experimenting with some axonometric or perspective sketches of the layout of the space you are creating (see pages 128 to 131).

Make a simple scale model of one or more of your ideas using thick card slotted together.

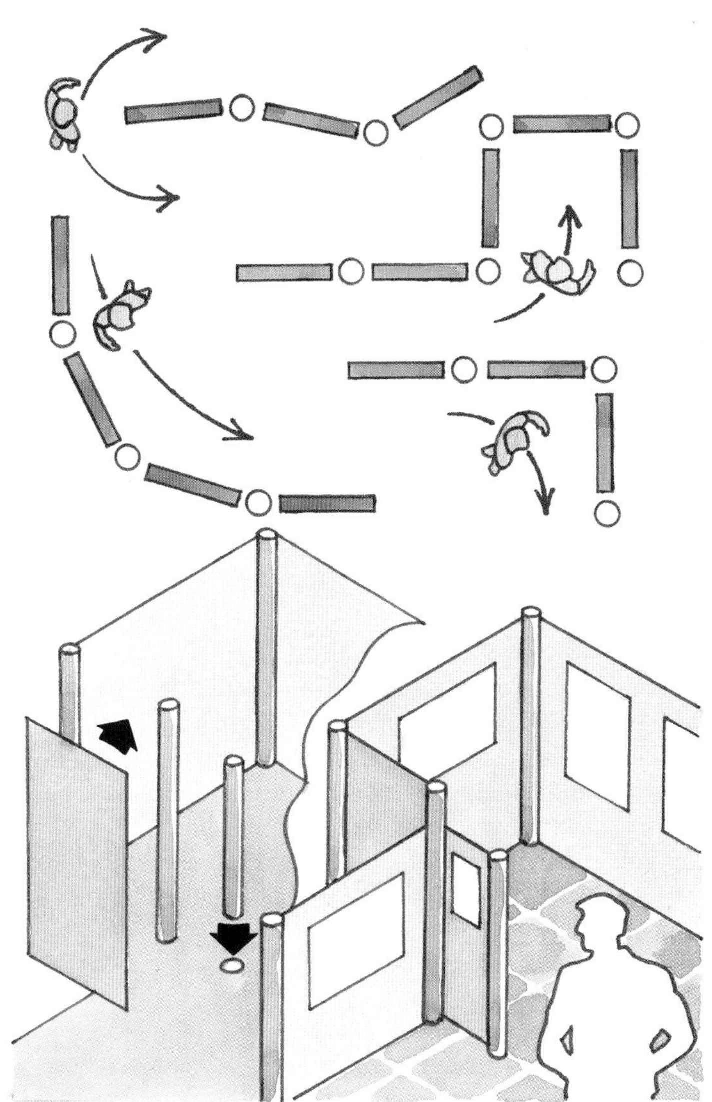

A Material Sample Board

Obtain some samples of real materials. These might include a selection of furnishing fabrics, curtains, carpets, wallpapers, paints and varnishes, wood veneers, etc. Some could be photographs taken from magazines.

Experiment by matching together different combinations and proportions of colours, patterns and textures of materials. You need to select samples for:

- ▷ the carpet
- ▷ different facings for the display boards
- ▷ curtains to hide the staff area.

Try to include some materials with unusual colours, patterns and textures which will help make the exhibition stand look distinctive.

Stick your final choice down onto a large sheet of card. You might want to complete several material sample boards and then decide which is the best one.

Architectural Drawings

Architectural drawings are made to visualise the design of buildings and structures. They are necessary in order for local authorities to consider and approve the building proposals. At a later stage they are used by architects to oversee building projects and builders to help guide them in the various work.

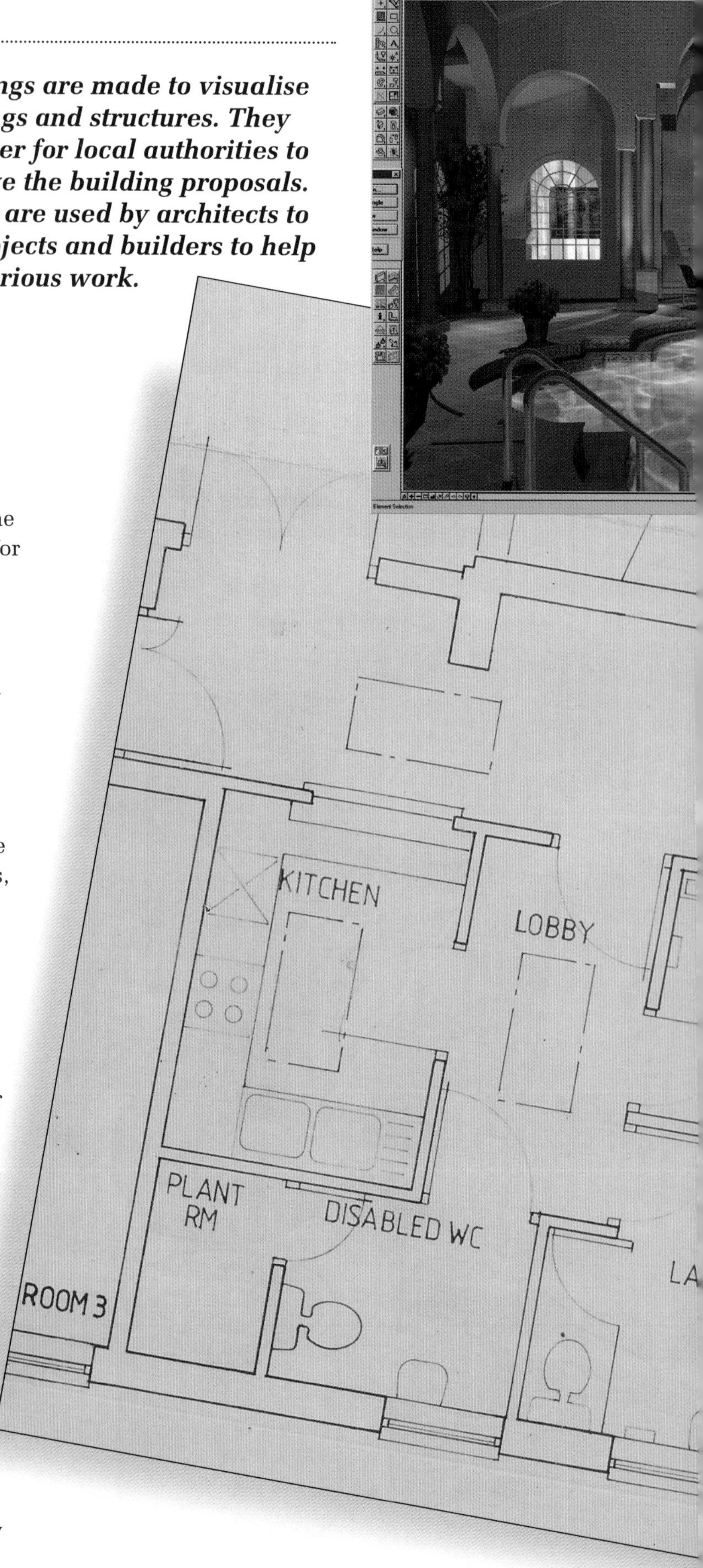

Plans

This is the general term given to all architectural drawings seen from above (i.e. a 'bird's eye view'). In the case of a building a separate drawing will be required for each floor. The exact shape and proportion of the building will be shown with all doorways, windows, internal partitions, stairways, and other main components included. The individual rooms will be marked and the layout of furniture and machinery may be shown.

Elevations

An elevation is a view from the side, front or rear of the proposed building. These, together with the plan views, will show the necessary information to allow the local authority to consider the planning application.

The elevations will show additional details such as height of windows, doors and type of roof. They give a clear indication as to the style and design of the building, and will usually indicate such details as brickwork, cladding, roof covering, and external timber work.

Elevations also lend themselves to colour work and the inclusion of people, planting, and vehicles to add realism. So called 'artist's impressions' may be done in this way even before the plans are drawn up. Usually it is a combination of plans, elevations and a site map that make up the planning submission to the local authority.

Architectural Symbols

Architects use a variety of graphic symbols to show details in their drawings. Many are dictated by convention and will be nationally and even universally

recognised. They will usually be drawn using special stencils or rub-down transfers. On a CAD system they can easily be retrieved from a library of symbols.

Axonometric Drawing

Designers often use 3D drawings to give a better idea of the environments they are creating, and to show additional information such as internal detailing, often in the form of a cut-away.

The most usual method is called an axonometric. A plan view is drawn as the basis of the drawing, tilted at 45 degrees. The vertical lines are projected from this to form the 3D image.

Before architectural plans can be drawn up a site survey is required and the client's needs considered. Consideration of the local environment is also important so that the new building will blend into its surroundings.

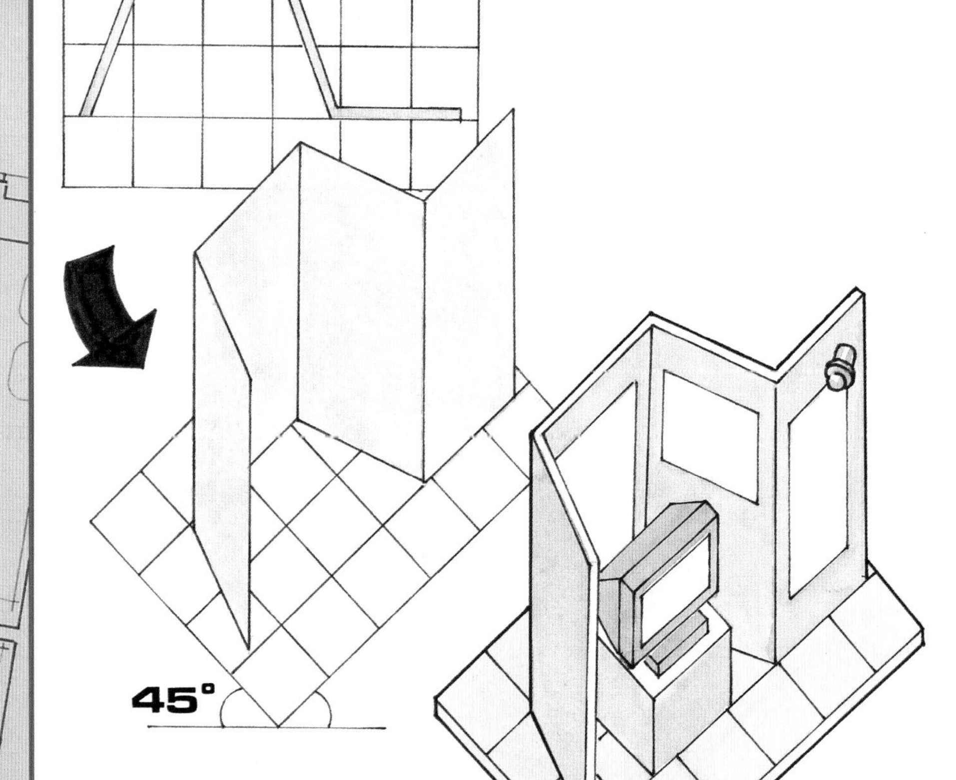

ACTIVITY

Identify an area of your school or home which contains two or three rooms connected by a passage-way of some sort.

Obtain, or draw out, a plan of the area you have selected. Include major items of furniture.

Develop a detailed axonometric drawing of the area. Use cut-away techniques to remove walls in the foreground to reveal the contents of the rooms.

Experiment by adding hatching and colour.

ICT

Most architectural plans and renderings are done using sophisticated CAD systems. In some packages it is possible to 'walk-round' a building.

IN YOUR PROJECT

- Use sketch axonometric drawings to help visualise spaces and places you are creating.
- Use measured axonometric drawings to help communicate your final design proposals.

In Perspective

Perspective drawing systems provide designers with methods for showing highly realistic visual interpretations of their ideas.

Simple freehand perspective drawings are quite easy to do. Complex, accurately measured, multiple-point perspective drawings take many years of practice and experience to get to look convincing.

All perspective is based on the principle that the lines on objects converge towards **vanishing points** in the distance, at which the lines meet. The vanishing points appear on a horizontal line at eye-level to the viewer. The placement of the vanishing points on the eye-level line gives different views of the object, and to some degree will be at the designer's discretion.

One of the clearest perspective images is the familiar view down a railway track where the principle of the converging lines may easily be seen and understood.

There are several types of perspective, each with different views that the designer may use.

One-point Perspective

One-point perspective is the most simple to construct with just a single vanishing point, which all left- and right-hand lines converge towards.

The object can be placed below the eye level (as in the example below) or above so that the underneath of the object can be seen, or breaking the eye level to give yet another view.

Single-point perspective is useful for simple views of objects and especially good for views of simple interiors such as exhibition spaces.

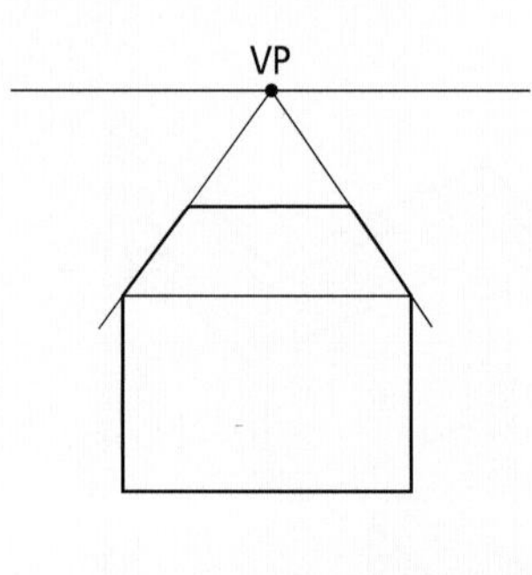

Two-point Perspective

Two-point perspective is an even more realistic form of three-dimensional drawing. The principles are much the same except that two separate vanishing points are used.

Taking the same example of the box below, one of the edges will appear closest to the viewer, having the effect that the box will be drawn at a slight angle.

All the lines except the vertical ones now recede to the appropriate vanishing points until the construction is complete. Two-point perspective is excellent for product design drawings and more detailed interiors such as kitchen layouts.

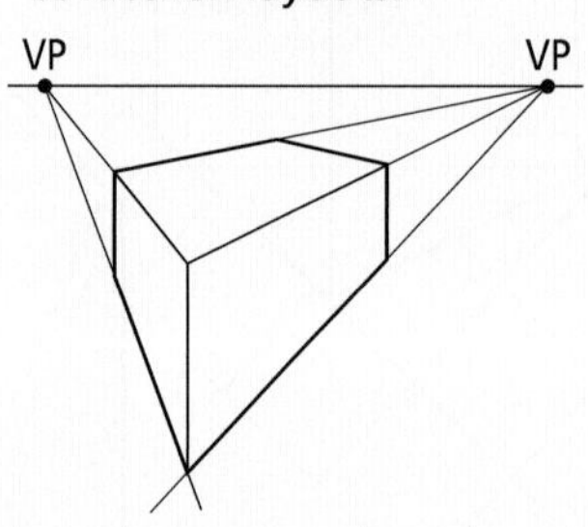

Three-point Perspective

Three-point perspective follows all the rules of two-point perspective, but in addition the previously vertical lines in the drawing now converge towards their own vanishing point which does not appear on the eye-level line. The effect tends to be that the drawing looks 'dynamic'.

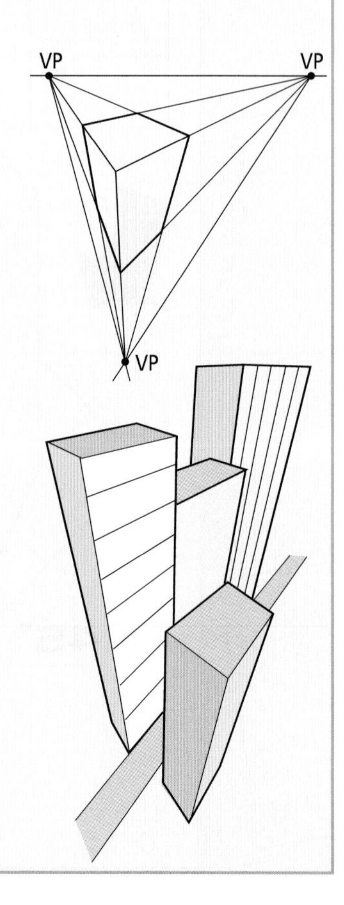

Construction Lines

Perspective drawings can either be in freehand sketch form, or they can be constructed more technically with the aid of drawing instruments. In either case, by tracing your original drawing any unwanted lines can be ignored and just the object can be reproduced ready for rendering and presentation.

Perspective grids are available to help construct more complicated drawings.

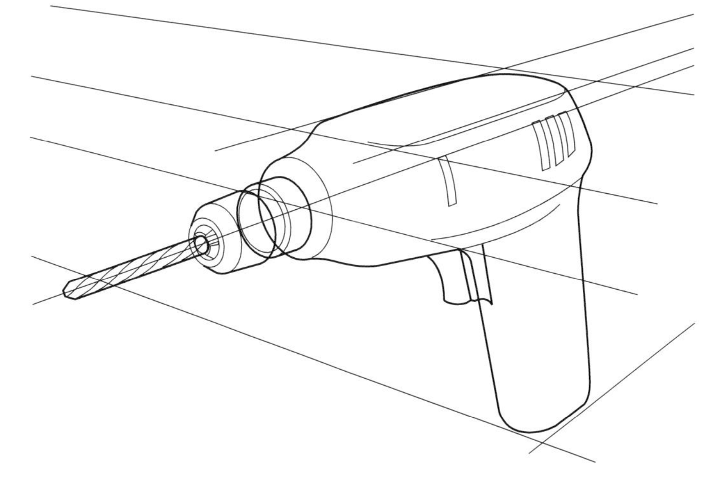

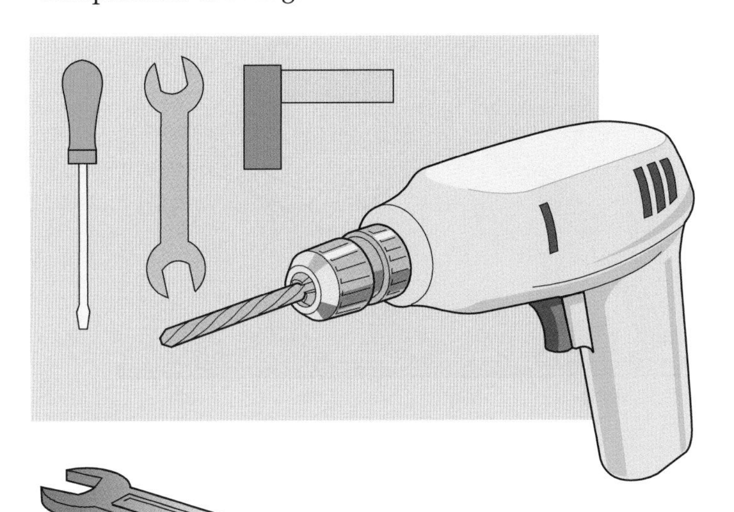

Tone and Colour

When adding tone or colour to a three-dimensional drawing that has a lot of 'depth', the colours should appear brighter in the foreground and duller the further away they are. This helps the three-dimensional quality of the drawing.

Drawing Circles

When drawing a circle in perspective it needs to be seen as an ellipse. A simple and quick method of visualising this is shown on the left:

1. Find the mid-points of each line.
2. Lightly draw a curve between the points.
3. Lightly sketch the ellipse until it looks correct.
4. Take care that sharp corners are not drawn.

When constructing a cylinder the same method can be used for each end (see left). Lines joining the edges of the ellipses can then be drawn to form the cylinder. These lines will conform to the usual rules of perspective.

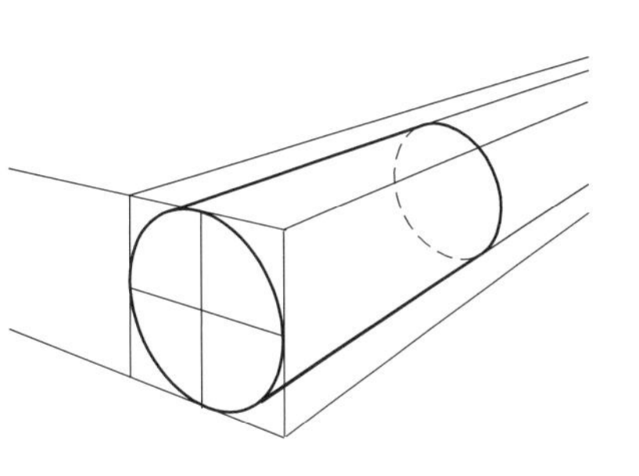

■ ACTIVITY

Identify a group of buildings, the interior of a room, or an object. Experiment by drawing your subject in one-, two- and three-point perspective.

Make notes on how long each drawing takes, how difficult it was, and how realistic the subject looks.

Add light and shadow and coloured rendering to your drawings.

IN YOUR PROJECT

- ▶ Try to use simple perspective sketches when developing ideas to help gain a more realistic idea of the objects and spaces you are creating.
- ▶ More complex, measured perspective drawings can be developed when preparing final presentation drawings to show a client, or possibly members of the public to help enable them to evaluate your proposals.

KEY POINTS

- One-point perspectives are effective at showing interior spaces.
- Two-point perspectives are effective at showing objects.
- Perspective drawings can prove to be very difficult and time consuming to prepare. A small inaccuracy can mean that the three-dimensional illusion fails to work.

Information Drawings

One of the most clear and effective ways of presenting information is to do it graphically. This is particularly true of numerical information. All information drawing should have good visual impact to attract the viewer's attention.

A computer graphics 'Draw' program is an excellent way of producing lively graphs, charts and diagrams.

Graphs

Graphs show the changing relationship between two factors, one of which is often time.

The upward or downward shape of the line or curve can be very quickly grasped.

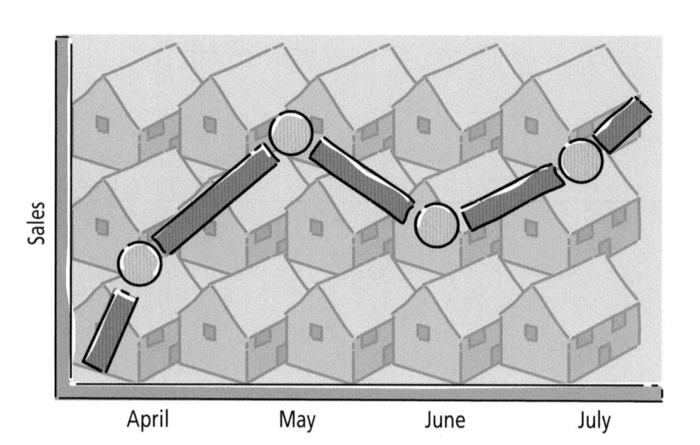

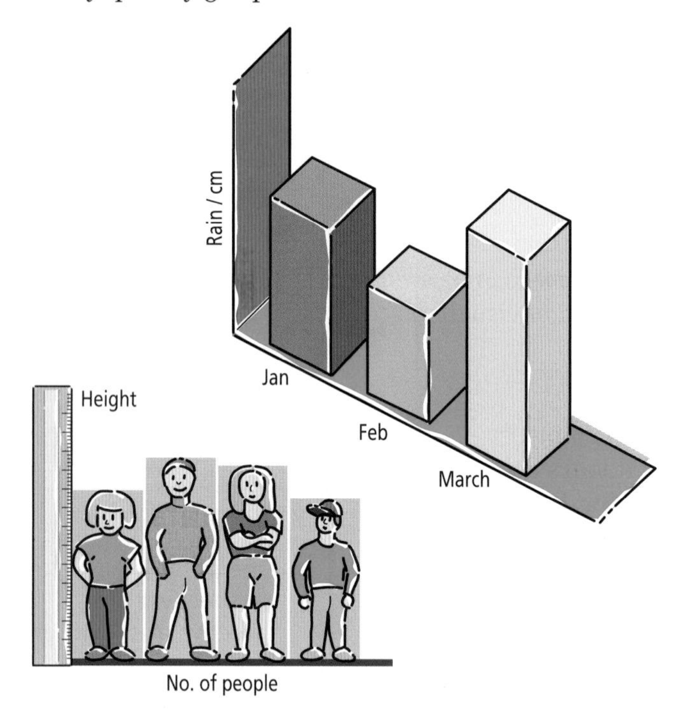

Bar Charts

The horizontal or vertical bars or columns of a bar chart are usually of the same width, but drawn at different heights. The bars can be sub-divided to show more complex information.

They may also be coloured, both solid or graduated, textured, or even contain images relevant to the subject matter. Another approach is to draw them as shapes or forms which relate to the subject of the chart.

Histograms are similar to bar charts, but with the bars or columns merged together.

Pie Charts

A pie chart is a diagram which uses a circle which is then cut into 'segments' to represent information as a proportion of the whole. The simple circle can be given much greater visual impact by adding colour and by the use of a third dimension. Showing a segment partly removed can also be very effective.

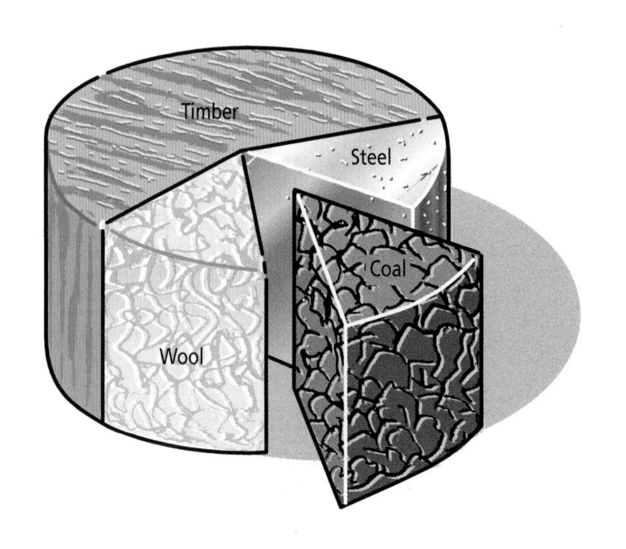

Signs and Symbols

Signs provide a way of communicating simple information quickly and forcefully. When they do not include any text they can be understood internationally.

Symbols are simplified images which represent the more complex shapes of objects or natural forms.

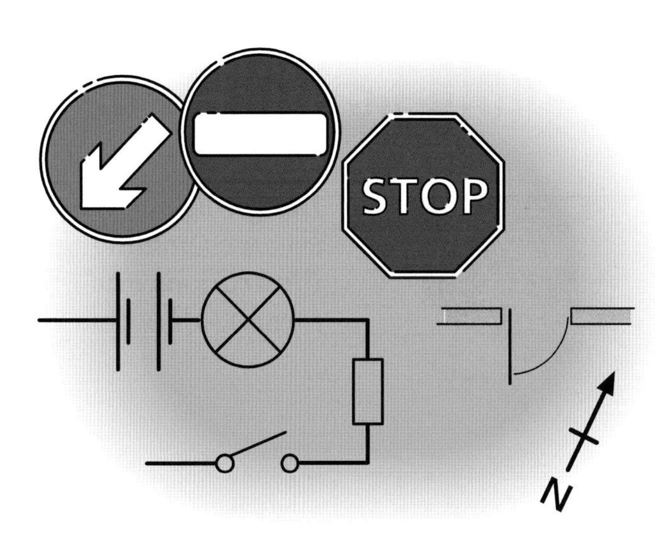

Pictograms and Ideograms

Pictograms and Ideograms are simple, logical 2D graphic illustrations that can be read in any language without translation. They are often used to indicate the location of public spaces (e.g, in leisure centres, shopping centres and transport terminals), on road-signs and on packaging.

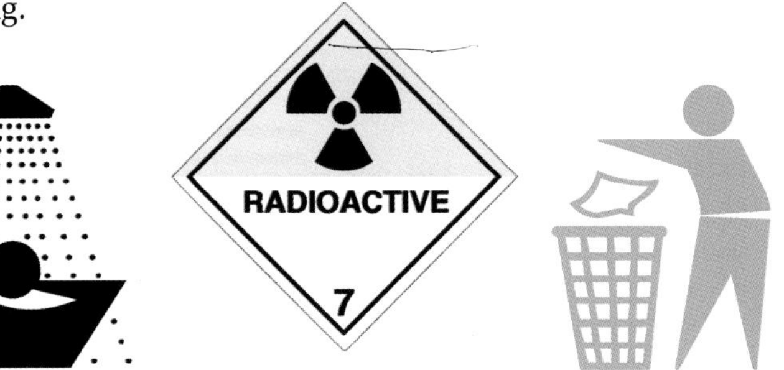

Flow Charts and Diagrams

Flow charts communicate sequences of events. There are usually drawn horizontally or vertically with lines linking the individual stages. Each stage can be represented as words and graphically. The distance shown between the stages can be made to represent physical space, or time.

A flow diagram is like a flow chart, except that it illustrates more of the parts of the system being represented.

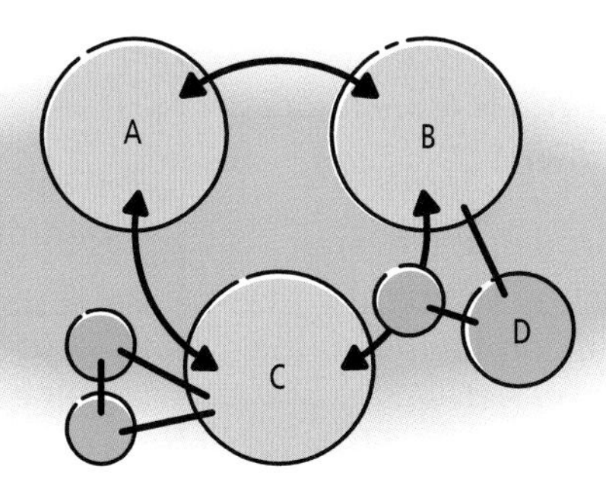

Expressive Arrows

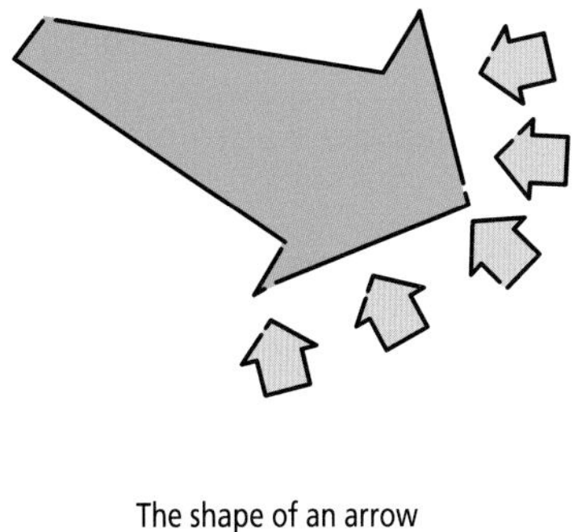

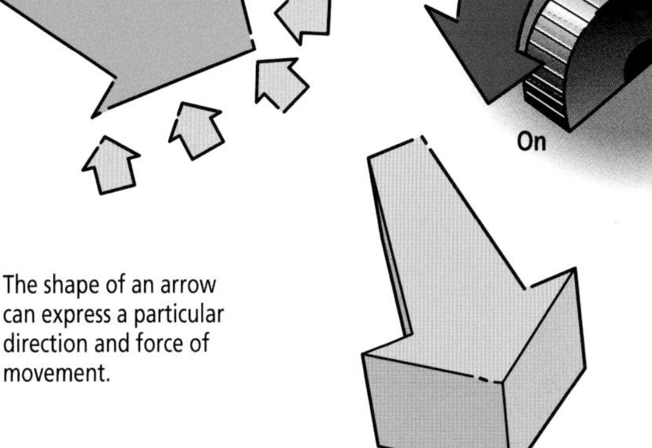

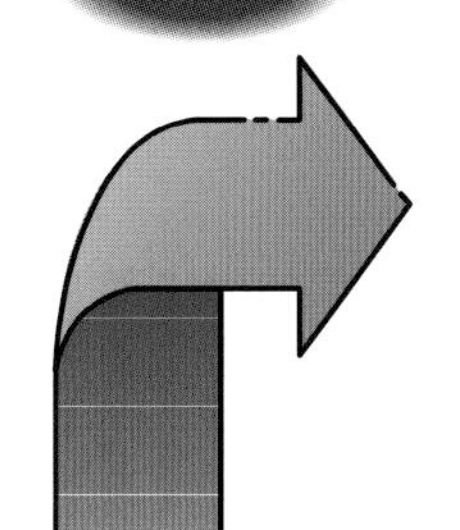

The shape of an arrow can express a particular direction and force of movement.

IN YOUR PROJECT

- Always be on the look-out for good opportunities to use a wide range of methods of graphic communication to illustrate information and ideas.
- Visual communication is essential in displays and exhibitions to attract interest as the audience will not have time to read detailed text.

Designing Leaflets and Brochures

Designing successful leaflets and brochures involves achieving the best layout of text, illustrations and graphic devices on the chosen format, all within the available budget.

■ ACTIVITY

Obtain a number of leaflets or brochures for domestic products, financial services, tourist attractions and/or hotels.

Analyse why they have been designed the way they have.

- ▷ How have different graphic designers solved similar problems?
- ▷ How effective do you think each is?
- ▷ How might they be improved?

Read through the design checklist on the right and base you comments on the points covered.

Present your findings with annotated illustrations.

Companies use brochures and leaflets to tell potential customers about the products and services they provide. First they need to attract attention to make sure they are noticed and picked up.

More detailed information is then provided to explain the particular benefits of the product or service, and how and where it can be obtained.

DESIGN CHECKLIST

Text (See also pages 38 to 41.)

What information is best provided in written form?

Which sections of text will be used as:
- ✓ titles and headings
- ✓ main (or 'body') text, and any 'small print'
- ✓ captions?

What size and style of typefaces will be most appropriate for each section?
- ✓ serif or sanserif
- ✓ traditional or decorative
- ✓ small or large
- ✓ thick or thin
- ✓ black, white, or coloured?

Illustrations

What information will be best provided by means of graphs and charts, plans, illustrative drawings, photographs, etc?

Will artwork be realistic, diagrammatic or impressionistic?

What size will illustrations be?

Will they be in colour or black and white?

What graphic devices, i.e. lines or flat areas of colour might be effective?

What text will illustrations need to be near to?

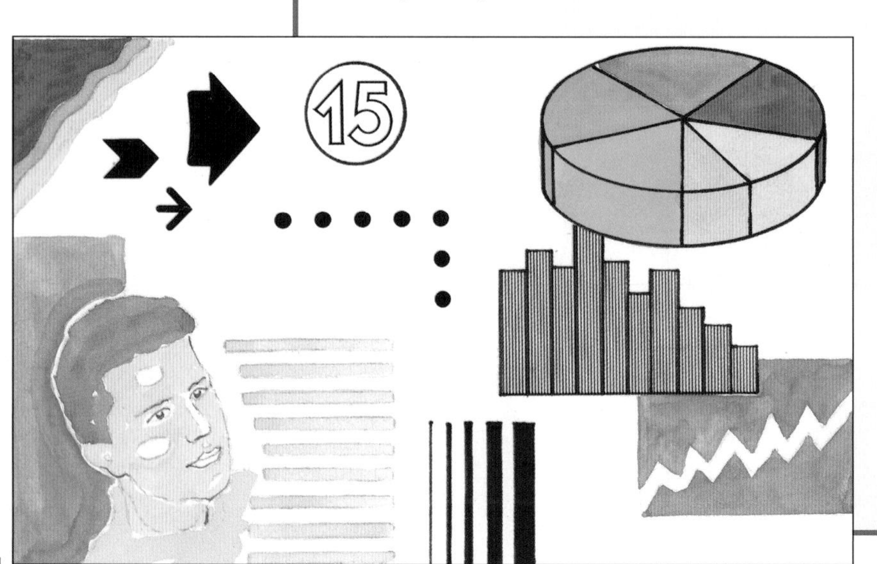

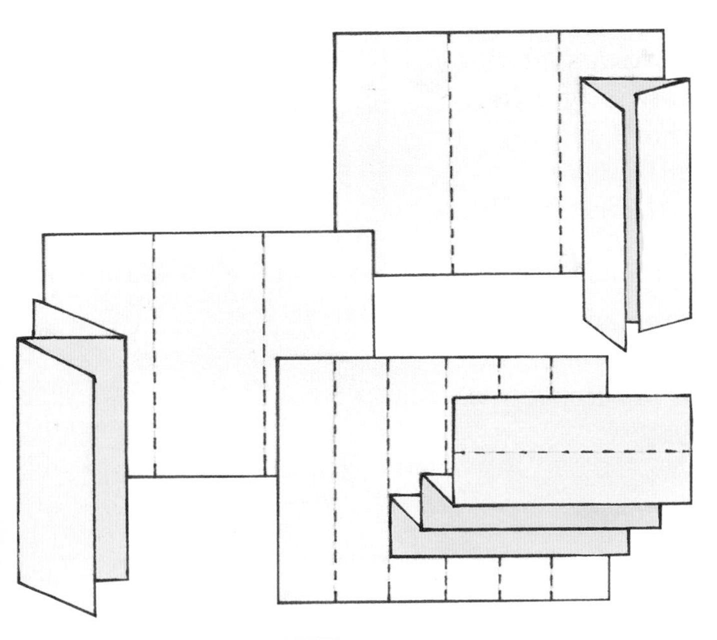

Format

What size and type of paper will be used, and how will it be folded and/or fastened together?

How will these decisions affect the printing and production costs?

(See also pages 76 to 77 on Printing Effects.)

Layout

How will the text and the illustrations be best arranged on the format?

Preparing a Presentation Rough

A presentation rough should be used to give the client a good idea of what the final printed product will look like. The printer will need a technical specification so that the text and images can be accurately set.

Making It

Plan out how to prepare the final artwork ready for the printers.

- ▷ What studio materials and facilities will be needed?
- ▷ How can DTP and CAD systems be used most effectively?
- ▷ What specialist jobs will need to be done by someone else?
- ▷ Which items need to be prepared first?
- ▷ Which can be dealt with later?
- ▷ What are the final finishing stages?

Costs of Production

Prepare a short report covering:

- ▷ how much the leaflets would cost to produce at different print-run lengths
- ▷ what printing processes will be used
- ▷ what checks on quality will be made?

Final Testing and Evaluation

Take a colour photocopy of your final artwork and make it up into a final dummy leaflet or brochure.

Show your final design to a number of people. Ask them to look through it, and then ask them some specific questions to discover how well they have understood:

- ▷ what the product or service being offered is
- ▷ its particular benefits
- ▷ where and how they can obtain the product or service.

Using Your Imagination

Imagination is an award-winning design and communications company renowned for its creativity. Clients such as BT, Ford, Disney, Lego and Cadbury commission them to communicate information about their products and services to a wide audience. To achieve this their range of work includes dramatic product launches, roadshows and special exhibitions, supported by printed brochures, film and video presentations and interactive multimedia computer experiences.

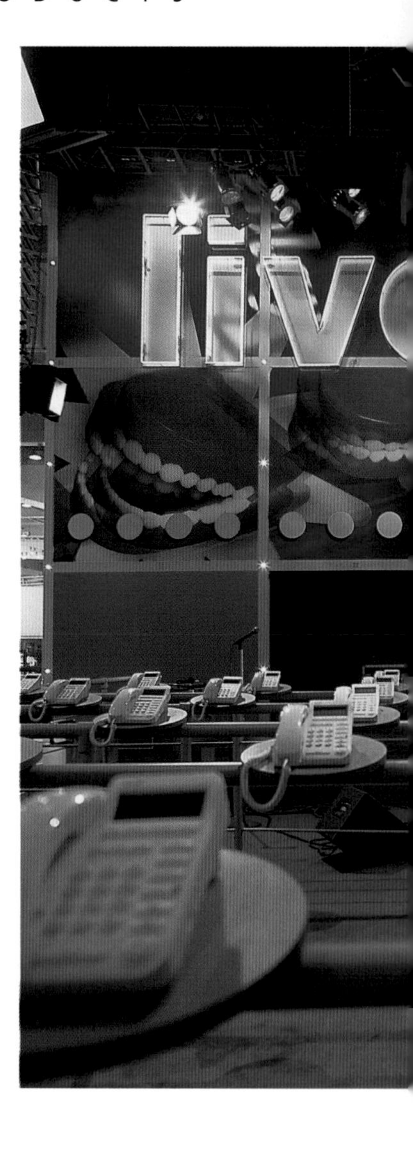

www.

To see some examples of well-designed websites go to:
www.BT.com
www.Ford.co.uk
www.Disney.com
www.LEGO.com
www.Cadbury.co.uk
NB: You may need to have the latest version of 'Flash' installed on your system in order to see the animated features of these sites.

Inside Imagination

Imagination's 200 strong staff includes architects, interior designers, CAD specialists, model makers, media relations experts, graphic designers, film and video directors and artists. Teamwork and total flexibility are essential ingredients to their success.

Ralph Ardill
Marketing Director

We meet impossible deadlines all the time. The only way you can do that is by being very flexible. If it needs doing, you do it, whether you're the cleaner or the project manager.

The company has an amoeba-like structure, always changing shape, depending on the nature of a particular job. The employees we recruit must bring energy, creativity, enthusiasm, and an ability to work with others to push their collective limits.

When work comes in, a brainstorm team quickly gathers. Project teams consist of appropriate specialists, but everyone is expected to be involved in every element of the task in an interdisciplinary approach.

Project teams decide how to organise themselves. Who joins the team, how they report to each other and how the work gets done is up to them. If some new piece of information comes to light, no one is afraid to change a decision the day after it's made.

Case Study: Launching BT at Live '95

BT invited Imagination to prepare proposals for its stand at Live '95. The theme of this annual event is electronic entertainment in the home. It provides an opportunity for major manufacturers to launch new products to the public, including some 180 000 enthusiastic young consumers – an important new market for BT's home entertainment products and services. BT had been at the event for the previous two years. However, research showed that visitors had little recall of their presence.

Whatever stand was created, it had several clearly defined tasks:

- to express key messages about BT's products and services
- to attract the target audience of young people, in competition with all the other stands
- to provide a memorable and enjoyable experience which would leave a lasting impression on visitors.

Entertainment Zone Ahead

Imagination's solution was to create the idea of the BT stand as an 'entertainment zone'. A lively, fast-moving show was presented at 15 minute intervals each day. When the show was not in progress, passing visitors were attracted by the chance to use the BT phones on the stand for free calls to anywhere in the UK. This helped maintain a constant audience – those using the telephones would be present for the start of each show.

The show was intended to be flexible, gathering crowds as it unfolded. When it started, overhead monitors around the stand and a video wall on stage came to life to catch the eye of passers-by. At the conclusion of the show, BT television commercials and information graphic sequences were relayed on the monitors.

■ ACTIVITY

Make a collection of current promotional material being produced by BT, or some other telecommunications company.

- ▷ What is their key message?
- ▷ What is their target audience?
- ▷ How are these reflected in their promotional material?

It's Good to Talk...

The stand was designed to capture the excitement and energy of live television studio set. A raised stage area was surmounted by overhead rigging which carried monitors, camera points and sound boons.

The presenters of the show interacted with visitors who were invited to appear on stage. BT staff mingled with the crowds, welcoming them, answering questions and giving details about BT products and services. There were games and competitions based on the theme of good communications, with prizes such as pagers and telephones. BT phonecards and merchandise were given out to the audience. In this way the atmosphere of the stand directly expressed BT's current brand message, 'It's good to talk'.

Research carried out immediately after the event was positive. It revealed that BT had made a lasting impression on 8% of visitors to Live '95 and a total of 70,000 people visited their stand making it the most popular at the exhibition.

Moving Images

Imagination are well known for creating high-quality corporate films which have the desired impact on the appropriate audience and are delivered within the agreed budget.

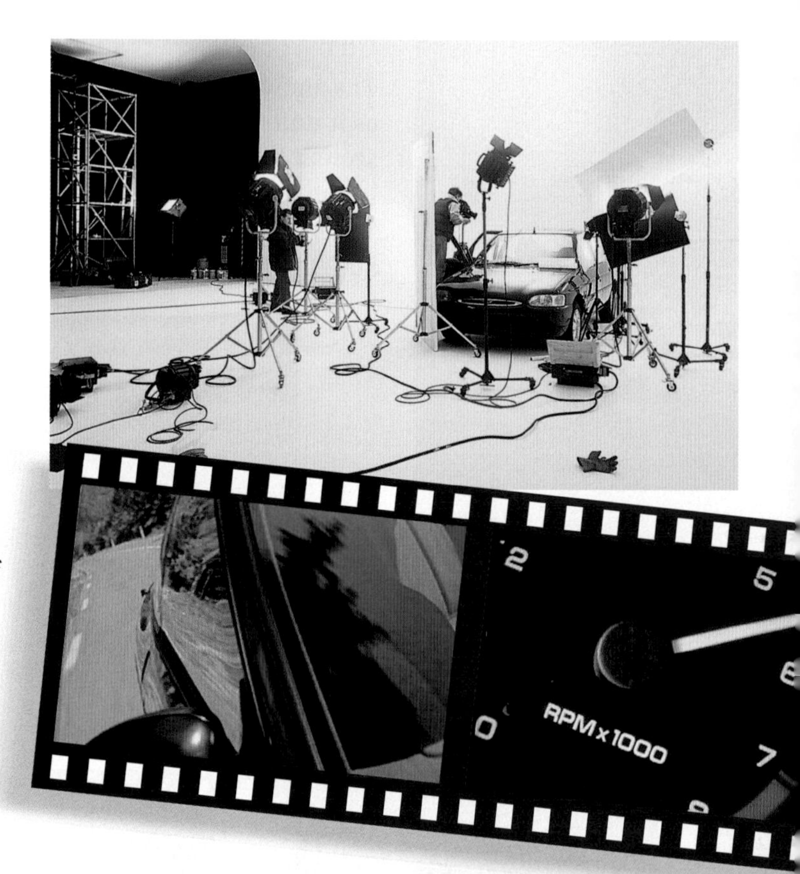

'Film' is a very general term which covers a range of different means of visual communication. One project may require a video, another a montage of still pictures, or perhaps an animated sequence.

Sometimes a project will require a complete range of integrated services. The launch of a new car, for example, might involve making:

- ▷ a television commercial showing the car in action
- ▷ a film presentation to alert the company's dealers
- ▷ a motivational film to inspire its manufacturing staff to make good quality products
- ▷ a promotional video for sales teams to highlight the key features of the new car
- ▷ a corporate presentation to explain to the press and public about the success of the company and the background to the design concept of the new car.

The first stage of creating a new promotional film or video is for the client and account director to get together to establish the brief and define the budget and the timescale.

The film director then works out a written outline, called the 'treatment'. This describes how the film will say what needs to be said and to which key markets and audiences it is directed. Detailed costings can then be calculated. An initial script will also be prepared – the words which will be spoken either on screen or by a voice-over, and a storyline which expresses the visual imagery of the proposed film.

The treatment, script and storyline will be presented to the client who may accept it, offer some suggestions or request some revisions.

Filming can take place in the studio or on location. Sets may have to be dressed and particular lighting systems and sound recording equipment obtained and set up before shooting can begin.

Once the script has been filmed, a rough edit is carried out. This is equivalent to a prototype or a mock-up, and shows all the scenes in the right order but without fades and wipes, sound effects and music. This stage is also shown to the client for approval, before the final version is completed.

As well as Imagination's own film and production crew, many other suppliers are needed, including equipment hire, film stock and processing laboratories, and post-production and duplication houses who make the final copies.

Presentation Graphics (1)

Imagination has a department dedicated to what it calls presentation graphics. Using the latest in CAD equipment and programs, the designers there work in five main areas: still imagery, logos, interactive multimedia, animation and virtual reality.

Still Imagery

This involves creating, manipulating and enhancing still images. These are often highly effective as part of presentations as the viewer tends to concentrate on detail within the picture, and so gain more focused information about the subject.

Logos

Here, 3D animated sequences are designed in which a company logo appears and comes together in some way. These are often used for opening titles for corporate films.

Interactive Multimedia Presentations

Conventional computer programs are usually limited to text and/or graphics. Multimedia can combine:

- written words or numbers (e.g. stories, poems, plays, data, formulae, etc.)
- graphic images (e.g. drawings, diagrams, animations, photographs, video clips, etc.)
- sound (e.g. voice, music, sound effects, etc.).

All these can be presented on screen at the same time. The quality of text, image and sound is as good as any book, photograph, TV programme or hi-fi system. The multimedia program can be controlled by a mouse and on-screen cursor, or by directly touching areas of special screens.

Every presentation graphics project starts with sorce key questions:

- Where will the presentation be seen?
- What audience will it address?
- What message must it send?
- What visual style will be most appropriate?

The way in which multimedia programs are created is different from the way in which TV programmes and books are written, and is commonly called 'authoring'. Special software programs exist to make linking the text, images and sounds easier.

To create successful multimedia designs you need to have:

- an awareness of the potential and limitations of the computer hardware and software
- an understanding of how to ensure that an interactive program will make sense to the viewer, and be easy and friendly to use
- the ability to work with different media: still photographs, film and video clips, computer-generated text and animations, and sound files.

Presentation Graphics (2)

Animation

Animation can create the appearance of complete (but as yet un-built) environments which the viewer 'walks' through on screen. The journey follows a pre-determined route which reveals particular features of the environment being travelled through.

ICT

- Multimedia provides exciting opportunities for communicating new design ideas to potential clients and customers.
- Virtual reality has enormous potential as a modelling tool for architects and designers. They will be able to electronically create and modify their buildings and products cheaply and quickly.

■ ACTIVITY

Find out if your school has any presentation software, and discover how it can be used. At a simple level you might be able to create a series of promotional 'slides' with text which will run automatically.

Virtual Reality

Virtual reality is an extension of multimedia. Environments, or 'worlds' can be experienced on a computer screen as an interactive 3D visual 'walk' or 'fly' around a computer-generated space. The viewer can decide where he or she wishes to turn or move. Alternatively the user can wear a stereoscopic videoscreen helmet and headphones to give a powerful feeling of being in the environment.

Imagination create 'quick-time virtual reality' (QTVR) sequences in which the viewer can choose to look upwards, downwards and around in a 360 degree circle from a fixed point, and zoom in and out for a closer or wider look. They have produced a QTVR movie of the interior of their London offices. It presents different areas through which one can wander, looking up, down and around, opening doors and going into rooms.

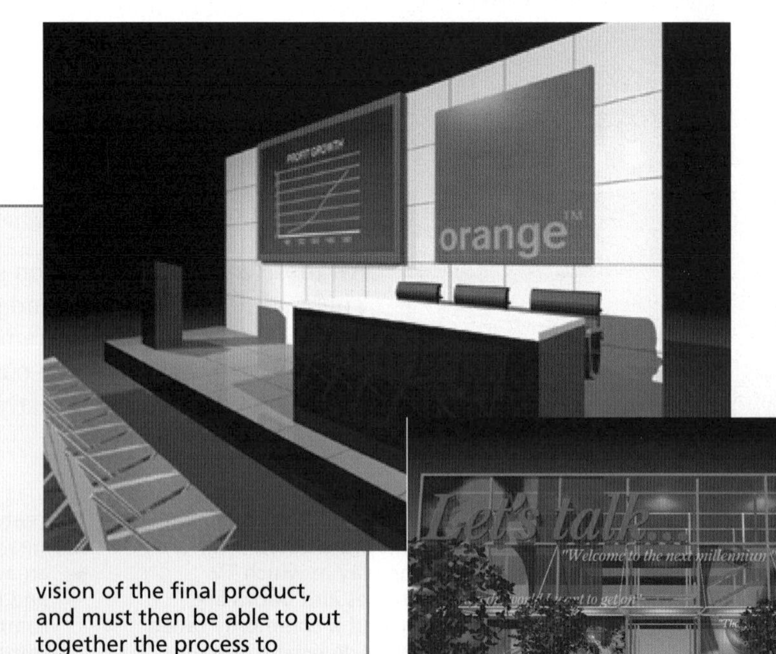

As a design company, we find that presentation graphics are highly effective in presenting ideas to clients.

For example, a 3D impression of a proposal for a new exhibition stand can be created on computer. Lighting and sound effects can be added in to give a better idea of the atmosphere. Clients can then be taken on a 'walk-through' of the design, providing a good indication of the way in which a visitor will ultimately experience the stand.

If the result was not what was intended or wanted, it is then possible to go back to the computer simulation and change the design as necessary, at a fraction of the time and cost of making complex scale models and full-size mock-ups in real life.

The skills needed to work in presentation graphics are wide ranging. It entails being adept at 3D modelling, video editing, graphic design and programming, together with a natural aptitude for problem solving. The designer needs to have a vision of the final product, and must then be able to put together the process to create it, using the most appropriate resources within the constraints of the budget and the needs of the client.

Into Cyberspace [1]

Cyberspace is the electronic world which links computers together. Using a modem and the Internet it is possible to send messages from one computer to another, anywhere in the world. These messages can be in the form of words, sounds and still or moving pictures.

Imagination uses the Internet to present its latest design work through a new communication medium which offers multimedia sound, vision and interactivity. This is often much more effective than printed materials in demonstrating what the company can do. The Internet offers a world-wide audience of potential clients who, if they wish, can get instant access to the company.

Imagination has a 'home' page. This is the first page you see when you visit their web site on the Internet. It welcomes the user and invites you to go on to read information about the company, or look through the Imagination 'on-line' magazine, which is called *Experience*.

The site is divided between company information and a less formal magazine section to appeal to different users, from existing and potential clients to designers and those simply interested in finding out more about the latest in design and multimedia presentation.

The corporate section outlines the philosophy, history of the company, its skills and resources, and provides a quick-time virtual reality tour of their London office, which is a highly original conversion of an Edwardian school.

Experience, meanwhile, has an opening contents page which explains what can be viewed. These are usually profiles of different departments within Imagination, case studies of some recent projects, and an opinion piece presenting the particular ideas of a member of the company or invited 'guest' contributor.

The whole site can be navigated page by page, or you can jump from page to page out of sequence using 'hot-spot' links. These are connections which instantly move you to another part of the site, or sometimes to a different site altogether.

Having a web site on the Internet also has the advantage of making it possible to keep a record of the number of users who visit the site. This enables Imagination to track which particular parts of the site have been visited the most. Research shows that users are attracted most by interactive elements such as video clips, short animations and sound files.

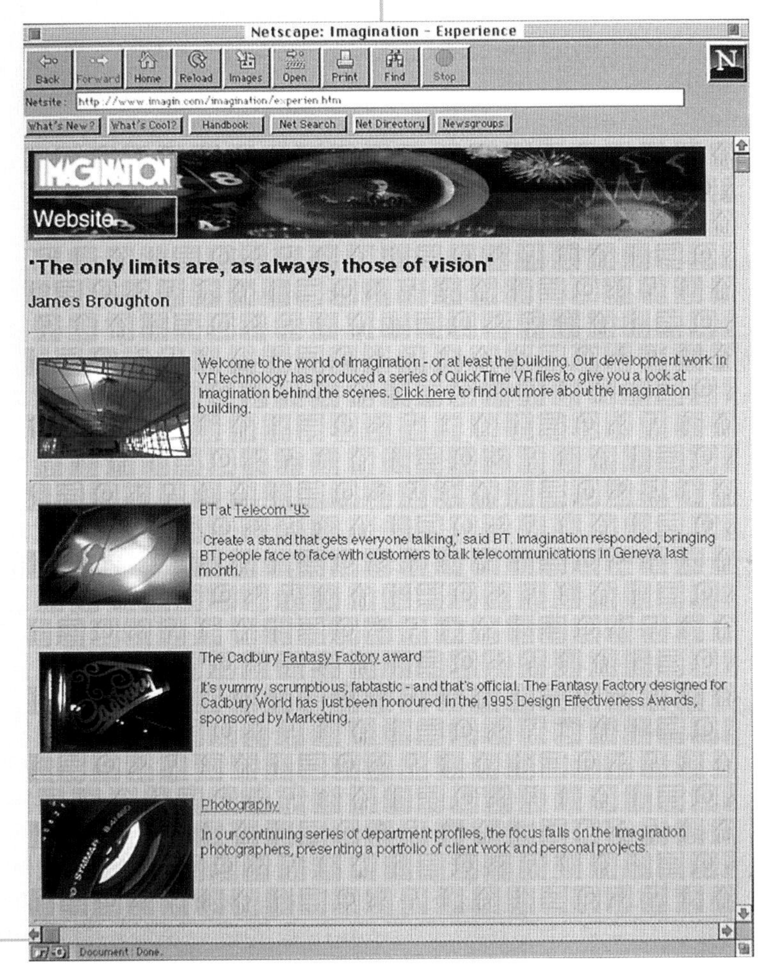

www.

Imagination was one of the first design and communication companies to have its own web site on the Internet. Its address is:
www.imagination.co.uk
To view some more examples of well designed web-sites, visit:
www.cartoonnetwork.co.uk
www.classicfm.co.uk
www.macromedia
www.momentumskate.com
www.sony.com
www.startdesign.com

You may need to have the latest version of 'Flash' installed on your system in order to see the animated features of these sites.

Into Cyberspace (2)

Designing a web site

Designing a page which is to be read on screen is different from designing one which will be printed. Pages can be as long or short as needed: there is no limit on the amount of text or images, except that the size and density of image affects the speed at which the page can be down-loaded onto the user's computer screen.

IN YOUR PROJECT

Could you design a web site as part of a promotional campaign to help launch a new product or service?

No one enjoys reading very small type on screen. The best images tend to be quite simple with strong colour contrasts. Everyone will see the pages in different ways. The size and resolution of monitors varies enormously. Some people prefer large rectangular windows on screen, while others opt for smaller, squarer frames.

Imagination designed their web site using a conventional format to be accessible to the user, pleasing to the eye, and instantly recognisable.

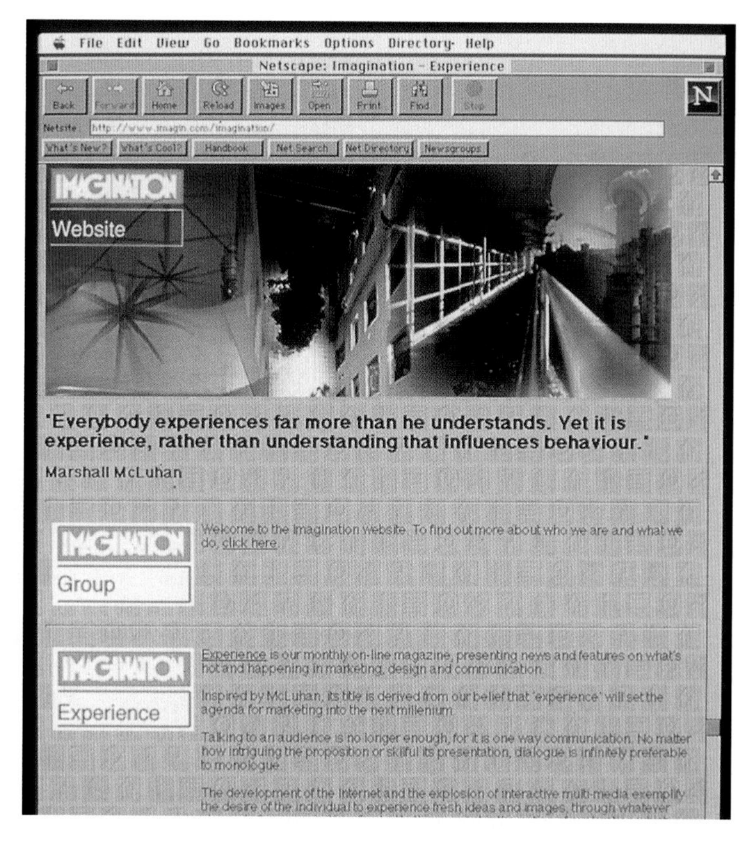

Imagining the Future

The capability of the latest information and communication technologies is immense. Our growing ability to communicate and receive complex messages is likely to bring about extraordinary and often unimaginable changes to the way in which we all live our lives. Many tasks which are currently done by people will be done quicker, safer, more efficiently and cheaper by sophisticated automated computer systems.

Verbal, aural and graphic communication – words, sounds and pictures – will play an increasingly key role in our future. There will be a great deal more information around that we can access – the problem will be knowing where to look, or where to place your message so that the people you want to will be able to find it.

As well as knowing how to use the new technology to create and receive our messages, we will also need to be able to tell the difference between well- and poorly-designed electronic graphic products. One thing that will never change however, no matter what media or technologies are used to communicate, will be the need for original and effective ideas. A sentiment that lies at the heart of Imagination's philosophy.

Planning and Making It

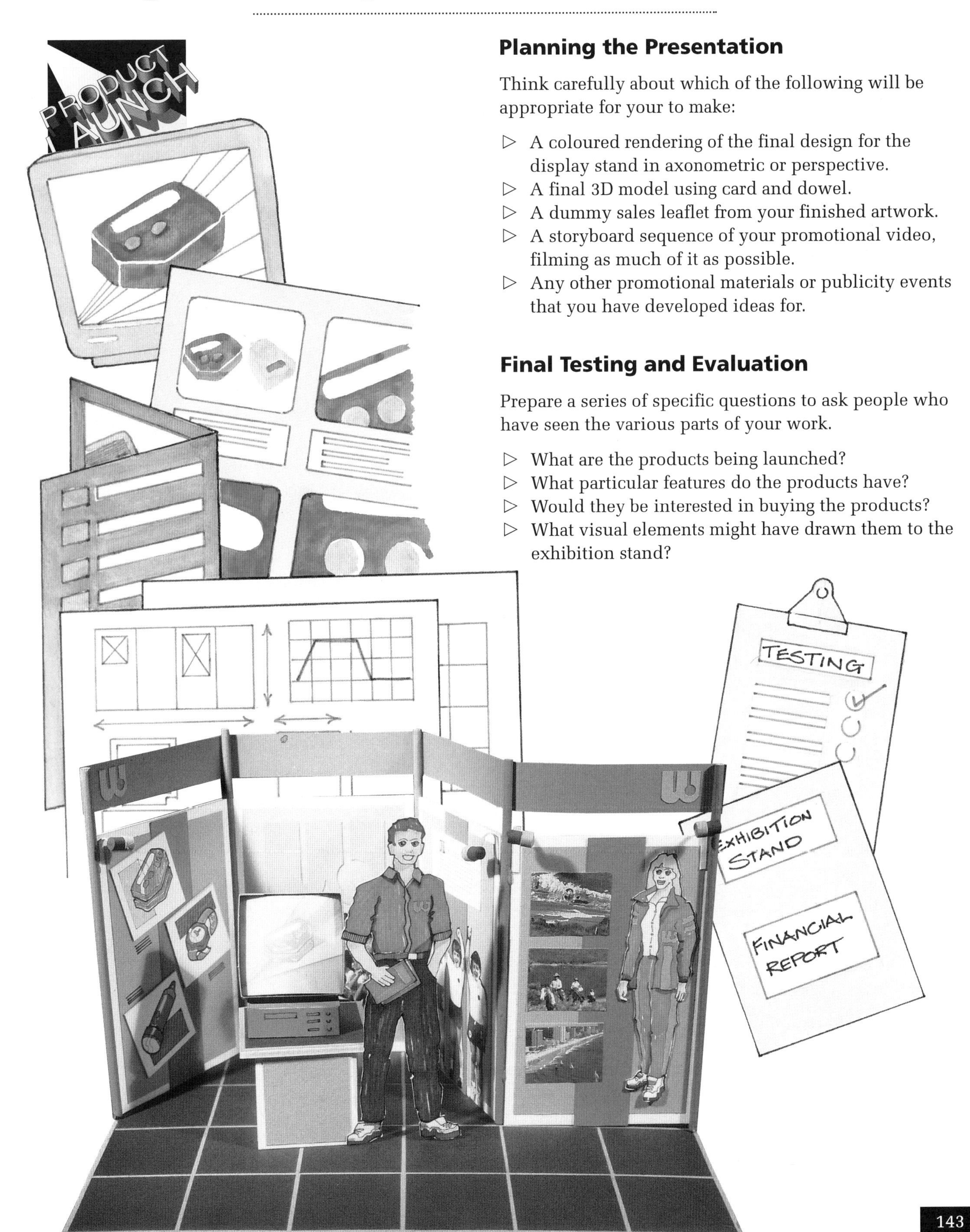

Planning the Presentation

Think carefully about which of the following will be appropriate for your to make:

- ▷ A coloured rendering of the final design for the display stand in axonometric or perspective.
- ▷ A final 3D model using card and dowel.
- ▷ A dummy sales leaflet from your finished artwork.
- ▷ A storyboard sequence of your promotional video, filming as much of it as possible.
- ▷ Any other promotional materials or publicity events that you have developed ideas for.

Final Testing and Evaluation

Prepare a series of specific questions to ask people who have seen the various parts of your work.

- ▷ What are the products being launched?
- ▷ What particular features do the products have?
- ▷ Would they be interested in buying the products?
- ▷ What visual elements might have drawn them to the exhibition stand?

Examination Questions

Before spending about two hours answering the following questions you will need to do some preliminary research into personal music systems.

To complete the paper you will need some plain A4 and A3 paper, basic drawing equipment, and colouring materials. You are reminded of the need for good English and clear presentation in your answers!

Make a study of the ways in which small electrical appliances are designed, displayed and packaged. Visit electrical retail stores and look at personal entertainment systems, e.g. radios, cassette, Minidisc & CD players, and MP3 recorders. Do not attempt to take the appliances or packaging apart!

Look at:

- the construction of the casing
- the arrangement of displays and controls
- the instructional leaflets
- the use of materials.

1. This question is about the graphic representation of data. *(Total 10 marks)*. See pages 132-133.

A survey of young people gave the following results for 'Thinking of buying a new personal music system'

Radio	=	10%
Cassette player	=	25%
Minidisc player	=	15%
Compact Disc player	=	20%
MP3 recorder	=	30%

a) Draw a coloured bar chart showing the results of the survey, label the parts and state the scale you used. *(8 marks)*

b) Give two reasons why you think MP3 systems are becoming more popular. *(2 marks)*

2. This question is about Designing. *(Total 26 marks)*. See pages 110-119.

A new UK home entertainment company called 'AudioCom' has developed a compact, portable MP3 audio recorder. It can be used to easily download music from the internet, and play it back through small earphones. The shape of the casing for the player now needs to be designed.

'AudioCom' has provided the following specification for the MP3:

- the size of the product must be no larger than 100mm in any one dimension
- the shape and colour must be distinctive
- the name of the company and the letters MP3 must appear on the casing
- the following displays and controls are to be included:
  Power on/off
  FF/Rew/Vol
  Operation status
- a belt clip must be provided

a) Using the information provided in the specification, produce a series of annotated sketches which show the development of your ideas for the MP3 casing. Make sure you show your development of ideas for:

(i) the overall shape of the product *(6 marks)*

(ii) the position of the controls and connection sockets, e.g. on/off, FF, Rew, volume controls, etc. *(8 marks)*

(iii) colours of the product and the position of the lettering. *(6 marks)*

b) How can a CAD package help you to design the casing? *(2 marks)*

c) How can a CAD package help you to present the final design? *(2 marks)*

As part of the design process you have to made a model (or mock up) of the MP3.

d) Why is a 3D model useful in the design process? *(2 marks)*

3. This question is about CAD-CAM. *(Total 20 marks)*. See pages 12-13, 118-120.

a) How can the final design for the MP3 player be transferred from a computer system to the miller or router that will manufacture the model body? *(2 marks)*

After the model has been designed on a CAD system it is to be formed out of solid material using a CNC miller.

b) Name a suitable material from which the model MP3 body can be machined. Give two reasons why this material us used. *(3 marks)*

c) What does CNC stand for? *(3 marks)*

d) The machining head of the miller moves in three directions, along three axis. Name these axis. *(3 marks)*

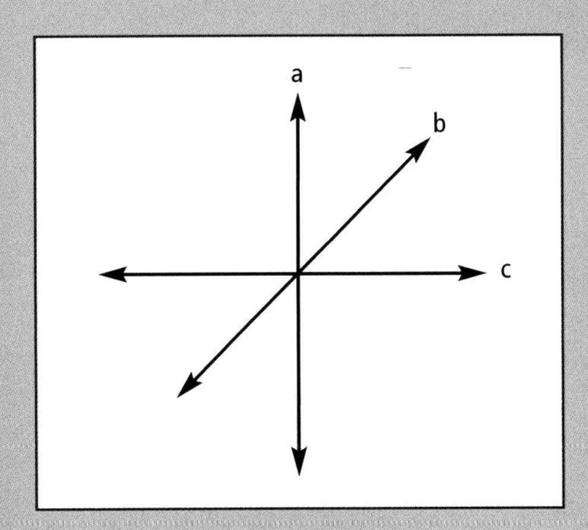

e) After the machining has ended, briefly describe the method of finishing the surface of the model. *(6 marks)*

f) State a safety precaution you would observe during the making of the model and say why it is necessary. *(3 marks)*

4. This question is about CAD/CAM. *(Total 12 marks)*. See pages 12-13, 118-120.

The company name 'AudioCom' and 'MP3' are to be made from adhesive vinyl and accurately fixed to the prepared model's body. Describe how this can done using a CAD/CAM system. Name all equipment and materials used. *(12 marks)*

5. This question is about Designing. *(Total 28 marks)*. See pages 106-111.

The MP3 player is to be sold in a card box. The top and the base must be secure.

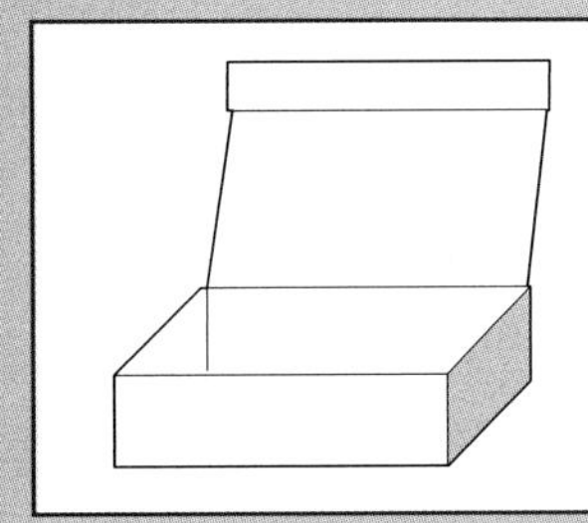

a) Give two reasons why the top and base must be secure. *(4 marks)*

b) Design a suitable base for the box which has secure interlocking flaps. *(6 marks)*

c) The top flop must be locked into position, design a mechanism which would prevent accidental opening. *(6 marks)*

d) Either:

(i) draw a neat freehand orthographic projection of the MP3 player. Use British Standard conventions show three important dimensions. *(12 marks)*

Or:

(ii) Produce a neat freehand pictorial sketch of the MP3 player. Use appropriate rendering techniques to represent the materials from which the player will be made when in production. State the type of sketching you have used. *(12 marks)*

6. This question is about communicating information graphically. *(Total 18 marks)*. See pages 132-133.

a) The inside of the top will have diagrams explaining how to use player. Draw thumbnail sketches showing the following steps

1. Log on to the net
2. Link to the MP3 music provider
3. Download the music to the hard disc
4. Copy the music file from the hard disc to the MP3 player
5. Enjoy the music

(10 marks)

b) The box has informational symbols printed on it. Some are legal requirements and others voluntary. Sketch one of each and explain its significance. *(8 marks)*

7. This question is about Processes and Manufacture. *(Total 6 marks)*. See page 81.

AudioCom have asked you to send your drawings and specifications for the MP3 player and box to a manufacturer.

Audiocom have to decide whether the first production run should be 10, 100 or 10,000 units.

List the four methods of production. State which method you recommend and why. *(6 marks)*

Total marks = 120

Around the World

BOARD GAMES
Design and make a board game and 3D playing pieces based on the book* Around the World in Eighty Days*, or some other classic novel you have read. You will also need to include an illustrated book of rules which are easy to understand. Finally, show how the game would be packaged and promoted. How could ICT be used?

PROJECT SUGGESTIONS

Product Rendering (page 112)
Information Drawings (page 132)
Packaging Materials (page 66)
2D CAD (page 42)

Investigation

Make a detailed study of a board game you have at home.

▷ What is the theme of the game?
▷ How is it played?
▷ What makes the game fun?

As well as your own thoughts, ask friends and other members of your family what they enjoy and dislike most about the game.

Think carefully about which books will supply you with enough visual imagery to base your ideas on.

Write a short summary of the plot of the book you decide to base the game on (about 100 words).

▷ List the main characters.
▷ List the main locations.
▷ List any important objects.

Remember to prepare a detailed design specification.

First Thoughts

Quickly sketch some ideas for a number of different types of game you could develop based on your book. Start by basing them on existing board games, such as Snakes and Ladders, Monopoly or Cluedo. For example, as with the Monopoly board, the players could move round landing on the capital cities which Fogg and his companions visit in, *Around The World in Eighty Days.*

Decide which your best idea is, and write down clearly why you think it will be the most successful.

Developing Your Initial Ideas

Think about your chosen approach in more depth. Sketch out alternative layouts for the board. In particular think about the following questions:

▷ How many people can play?
▷ Will players need to throw dice?
▷ Will they pick up chance cards?
▷ Can they decide which direction to move in?
▷ Can other players' moves be blocked?
▷ Are points scored on the way?
▷ How does a player finally win?

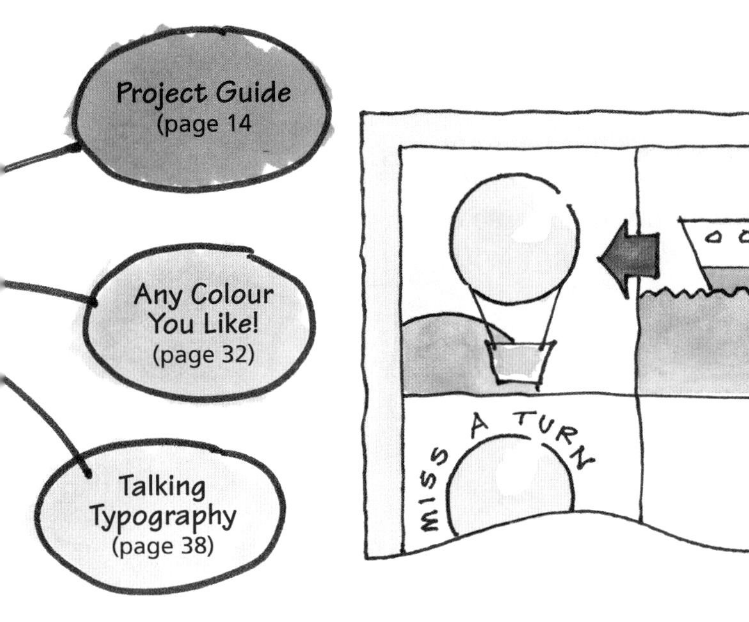

Developing the Graphics

Aim to make the board as visually interesting as possible. Use images and colours to help set the scene of the book your game is based on.

Developing the Rules

It is important to make the rules as easy to follow as possible. They will need to be illustrated to help identify playing spaces and pieces and how they can move.

Developing the Packaging

How and where will all the parts be housed in the box? What will the outside of the box look like?

Remember the packaging graphics play a very important role in promoting sales.

Making and Testing a Prototype

Think carefully about some aspect of your game that you need to test out to see how well it works, such as:

- ▷ how quickly it gets started
- ▷ whether it becomes dull or repetitive at any particular stage
- ▷ whether it is too obvious who will win from early on.

Write down exactly what you want to test out.

Make a drawing of your game which is good enough for people to play. It need not be very neat and should not contain detailed colour graphic work at this stage.

Test your game out by getting a group of people to play it. Make a careful note of what happens. As a result of what you discover, what are you going to need to change or work further on?

Making It

Carefully plan the making of the final versions of the board, playing pieces, rules and packaging. Make everything to the highest possible quality. Use any opportunities you have to use computer-graphic and DTP programs, colour photocopiers, etc.

Final Testing and Evaluation

Present your final realisation to a group of children.

- ▷ How quickly can they work out the rules?
- ▷ Do they find the game interesting and enjoyable to play?
- ▷ How might it be improved?

Remember to discuss the process of designing and making you used referring to the comments made by the children.

Simply Sundaes

A new chain of high-street ice-cream parlours, Simply Sundaes, is planning to open 50 outlets across the country. As well as a wide range of specialised ice-cream-based dishes they will also serve other desserts, as well as a range of teas, coffees, soft drinks, etc. How could ICT be used?

Architectural Drawings (page 128)
PROJECT SUGGESTIONS
Lifestyle Images (page 96)
Printing Effects (page 76)

The Task

You have been asked to:

▷ Identify a possible site in your locality for them to open a parlour, giving reasons for your choice.
▷ Develop a visual identity for the whole nationwide chain and show how it would be applied to your local site. The graphics you design should be applied to a range of items such as the shop-front and sign, interior design, the pop-up menu graphics, staff uniforms and advertising and promotional materials. Where possible, indicate the projected cost of batch producing these designs for all 50 outlets.

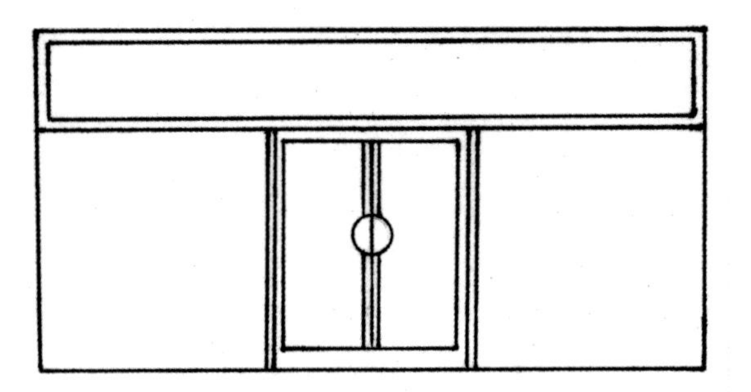

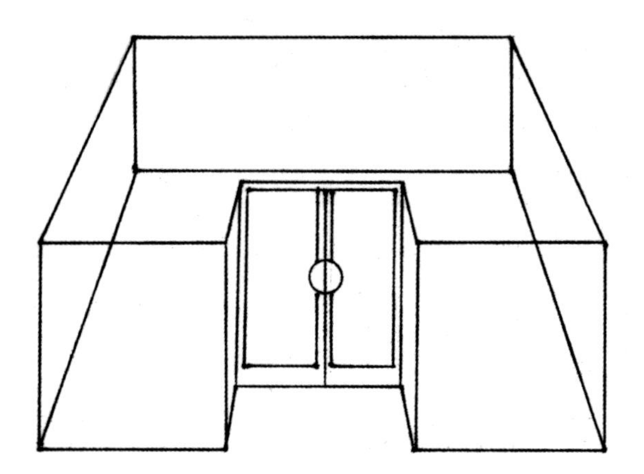

Investigation

Explore your local neighbourhood or a nearby town to find some possible sites for the parlour. Undertake a site study of the most suitable location (i.e. make sketches, take photographs, find and draw maps, etc.).

▷ Is the size right?
▷ Is it in a good position to attract the public?
▷ Is there any parking space?
▷ Is anyone likely to complain?

Present your findings on an A2 display panel.

Brainstorm and choose a visual theme for the chain. If you wish you may suggest alternative names to Simply Sundaes to tie in better with your theme. Make a collection of pictures cut from magazines, photocopies and sketches which relate to your theme.

Develop and undertake a small survey to find out more about the sort of people who might visit the parlour and their favourite ice-creams and other desserts.

Design Specification

Write a clear specification for the visual identity you intend to create, and its application to a range of items.

Project Guide (page 14)

Graphic Identity (page 46)

Into Print (page 72)

Making and Testing Prototypes

Prepare mock-ups of your designs. Show them to a group of people to get their responses. What product is being sold? What qualities of product and service might be expected? What sort of prices might be charged? How well does it meet other specific requirements of your original specification?

Prepare your questions carefully beforehand. Think about what sort of information you need to learn about your ideas. Record your results. What changes do you plan to make?

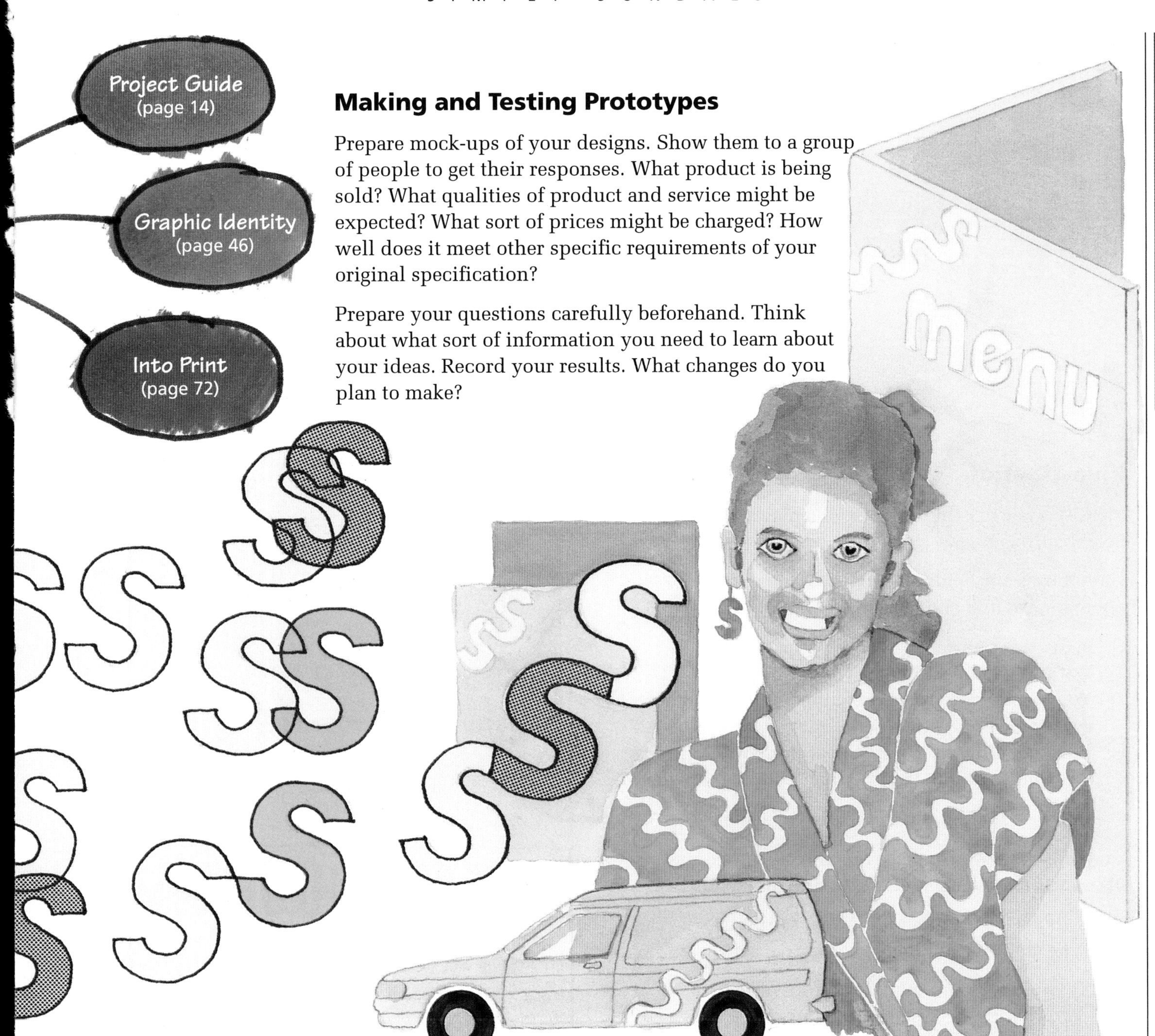

Making It

Imagine you are going to present your ideas to the management team of Simply Sundaes. What are the key aspects of your designs they need to know about? How can you communicate these most effectively?

Plan your time carefully to ensure you finish on time.

Communication methods
- Display-panels?
- Mock-ups?
- Scale models?
- Technical and financial reports?
- Graphs and charts?
- Video or computer-generated presentations?

Final Testing and Evaluation

Show your final work to a group of people. Ask similar questions to those used when testing the prototypes. Have the problems been sorted out? Is it a better solution?

Further ideas
- Which famous personality would you invite to open the ice-cream parlour?
- What special offers could be made to encourage people to visit during the first week?
- How could the opening be publicised on local radio and newspapers?

Fun Machines

Architectural Drawings (page 128)

Presentation Matters (page 121)

Orthographic Drawings (page 106)

Your County Council wants to develop its provision of play areas and playground equipment for the under-sevens across its regions. It is looking for a modular design of play structure which can be constructed in different sizes and formations at about 100 different sites. You have been invited to submit design ideas for such a play structure.

Investigation

Identify some existing and possible play areas in your neighbourhood.

Undertake a site study (i.e. make sketches, take photographs, find and draw maps, etc.).

- ▷ Is the size right?
- ▷ Is it in a good position for young children?
- ▷ What is the existing play equipment like, and how is it used?
- ▷ What condition is the area in at present?
- ▷ What other nearby facilities and buildings are there?

Present your findings on an A2 display panel.

Undertake a survey to discover more about the sort of play areas that young children and their parents like.

You will need to find out more about the range of sizes of young children and their physical capabilities.

Where could you find out the appropriate safety standards for playground equipment?

See if you can obtain and analyse some leaflets and brochures which describe existing play structure systems.

Design Specification

Write a clear specification for the system of play equipment you need to design. Make sure you make a series of statements about things such as safety, the physical needs of children and adults, visual appeal and appropriateness for its surroundings, maintenance and upkeep, and costs of production.

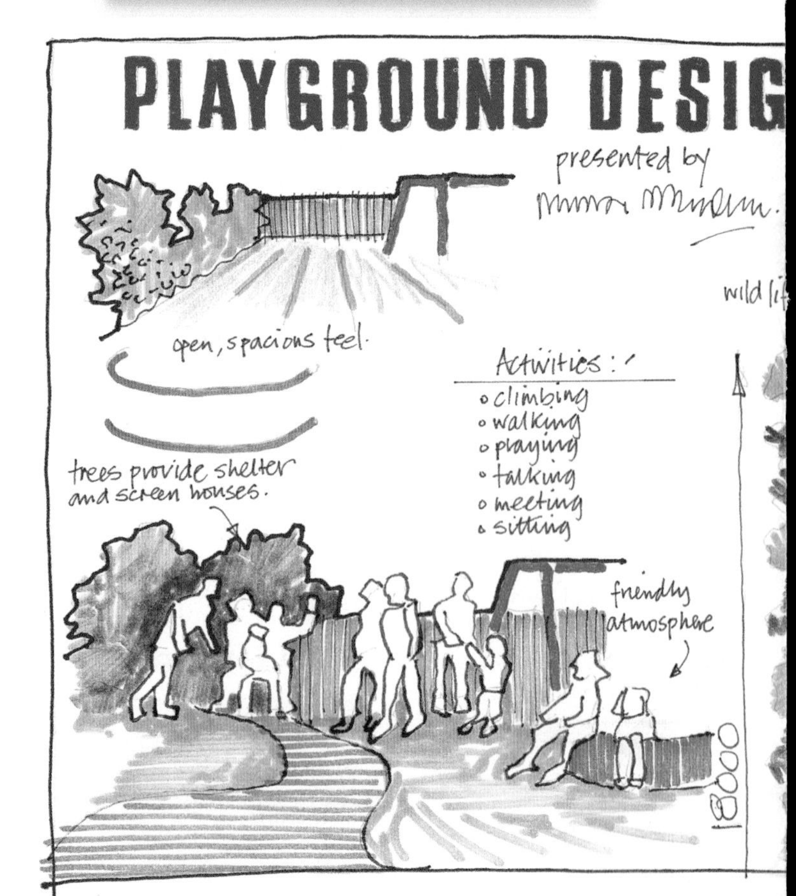

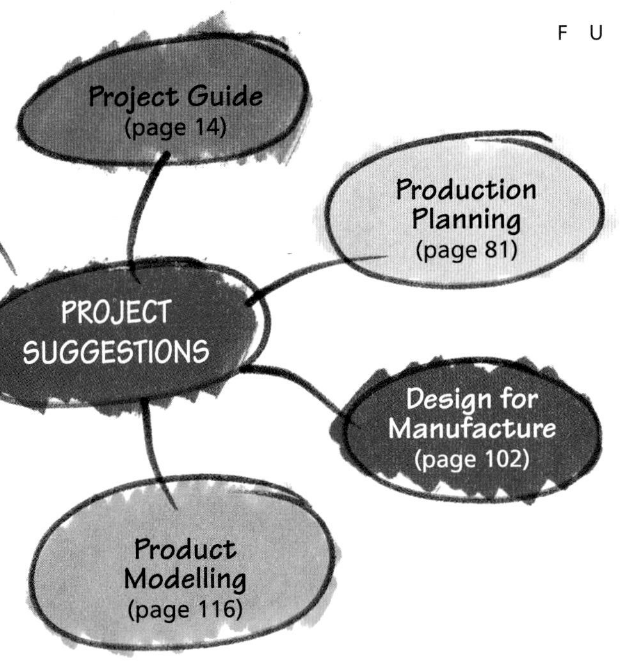

First Thoughts

You may wish to base your initial ideas on existing designs. Experiment by changing the shapes, colours, sizes, materials, methods of fastening. Try to develop a strong visual appearance which will be distinctive and quickly recognised as a well-designed play area.

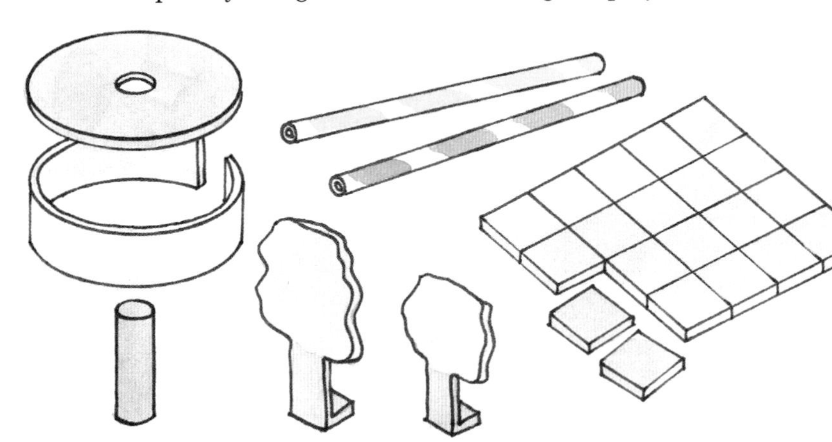

Making and Testing Prototypes

Prepare mock-ups of your designs. Show them to a group of children and parents to get their responses. How well does it meet other specific requirements of your original specification?

Prepare your questions carefully beforehand. Think about what sort of information you need to learn about your ideas. Record your results. What changes do you plan to make?

Making It

Imagine you are going to present your ideas to the County Council. What are the key aspects of your designs they need to know about? How can you communicate these most effectively?

- ▷ Display panels?
- ▷ Mock-ups?
- ▷ Scale models?
- ▷ Technical and financial reports?
- ▷ Graphs and charts?
- ▷ Video or computer-generated presentations?

Plan your work carefully to ensure you finish on time.

Final Testing and Evaluation

Show your final work to a group of children and parents, and if possible someone from your local council. Ask similar questions to those used when testing the prototypes. Have the problems been sorted out? Is it a better solution?

Index